Cryogenic Super-Resolved Fluorescence Microscopy

Der Naturwissenschaftlichen Fakultät
der Friedrich-Alexander-Universität
Erlangen-Nürnberg

zur

Erlangung des Doktorgrades Dr. rer. nat.

vorgelegt von

Siegfried Weisenburger
aus Stuttgart

Bibliographic information published by the Deutsche Nationalbibliothek

The Deutsche Nationalbibliothek lists this publication in the Deutsche Nationalbibliografie; detailed bibliographic data are available on the Internet at http://dnb.d-nb.de .

ISBN 978-3-8325-4364-8

Logos Verlag Berlin GmbH
Comeniushof, Gubener Str. 47,
10243 Berlin
Tel.: +49 (0)30 42 85 10 90
Fax: +49 (0)30 42 85 10 92
INTERNET: http://www.logos-verlag.de

Als Dissertation genehmigt
von der Naturwissenschaftlichen Fakultät
der Friedrich-Alexander-Universität Erlangen-Nürnberg

Tag der mündlichen Prüfung: 04. Oktober 2016

Vorsitzender des Promotionsorgans: Prof. Dr. Georg Kreimer

Erstberichterstatter: Prof. Dr. Vahid Sandoghdar
Zweitberichterstatter: Prof. Dr. Stefan W. Hell
Drittberichterstatter: Prof. Dr. Jens Michaelis

"And by the help of Microscopes, there is nothing so small, as to escape our inquiry; hence there is a new visible World discovered to the understanding."

—Robert Hooke, *Micrographia* (1665).

Summary

The significance of super-resolved fluorescence microscopy beyond the diffraction barrier was recognized by the Nobel Prize in Chemistry in 2014. One of these methods is based on pinpointing the position of single fluorescent molecules, in which the center of the point-spread function can be determined with arbitrary localization precision depending on the available signal-to-noise ratio. At room temperature, the signal of a fluorophore is limited by photobleaching resulting in a typical resolution on the order of twenty nanometers. Such super-resolution microscopy methods already allowed for resolving sub-cellular structures and organelles, and are starting to enable discoveries in neuroscience, molecular biology and other life sciences. One can dream about increasing the optical resolution by another two orders of magnitude in order to directly resolve sub-molecular structures such as constituents of molecular complexes or even protein structure itself. The aim of the present thesis is to accomplish exactly that.

In chapter 1 we give a critical overview of various recent developments in optical microscopy and put the present work into context. In addition to the popular super-resolution fluorescence methods, we also discuss the prospects of other techniques and imaging contrasts, and consider some of the fundamental and practical challenges that lie ahead. Next, chapter 2 describes the cryogenic and optical experimental setups which were constructed within the framework of this thesis. In chapter 3 we present a survey of the photophysics of organic dye molecules at cryogenic temperature; we also introduce the experimental techniques used in this work and discuss some photophysical effects which we observed at low temperature like dipole-dipole coupling between fluorophores. Chapter 4 then gives a summary of the mathematical framework of the single-molecule data analysis as well as the concrete implementation of the algorithms.

Chapter 6 presents our results from cryogenic super-resolution imaging of whole cells. Measurements on frozen CHO cells stably expressing dopamine receptors allowed us to determine the distance between the protomers of receptor homodimers. In chapter 5 we then introduce our cryogenic localization microscopy method which allows for single-molecule localization at the Ångström level due to the substantial improvement of the molecular photostability at liquid helium temperature. We furthermore verify the feasibility of co-localization and cryogenic distance measurements by resolving two fluorophores on the backbone of a double-stranded DNA at nanometer separation (chapter 7). In chapter 8 we then present our results on resolving the positions of multiple fluorophores attached to a protein using cryogenic localization microscopy. By applying algorithms borrowed from cryogenic electron microscopy, we can reconstruct a three-dimensional density map for the positions of the fluorescent labels with a resolution of a few Ångström, yielding excellent agreement with the expected crystal structure. Finally, in chapter 9, we suggest and discuss possible technical improvements, extensions and applications of the method presented in this work.

The here presented technique allows us to gain structural information of molecules that might not be accessible via existing analytical methods such as X-ray crystallography, cryogenic electron microscopy or magnetic resonance spectroscopy. These results mark record resolution and demonstrate that optical resolution can be pushed beyond the diffraction limit by nearly one thousand times.

Zusammenfassung (German)

Die Bedeutung superauflösender Fluoreszenzmikroskopie jenseits der Beugungsgrenze wurde 2014 durch den Nobelpreis in Chemie anerkannt. Eine dieser Methoden beruht auf der Positionsbestimmung einzelner fluoreszenter Moleküle, wobei das Zentrum der Punktspreizfunktion mit beliebiger Lokalisierungspräzision bestimmt werden kann, abhängig von dem zur Verfügung stehenden Signal-Rausch-Verhältnis. Bei Raumtemperatur ist das Fluoreszenzsignal eines solchen Markers durch Photobleichen begrenzt, was eine typische Auflösung im Bereich von etwa zwanzig Nanometern zur Folge hat. Solche superauflösenden Mikroskopiemethoden ermöglichten schon das Auflösen von subzellulären Strukturen und Organellen und beginnen nun Entdeckungen in den Neurowissenschaften, der Molekularbiologie und anderen Lebenswissenschaften möglich zu machen. Man kann davon träumen, die optische Auflösung um zwei weitere Größenordnungen zu verbessern, um submolekulare Strukturen wie etwa Teile von Molekülkomplexen oder gar Proteinstruktur direkt aufzulösen. Das Ziel der vorliegenden Arbeit ist es, genau dies zu erreichen.

In Kapitel 1 wird ein kritischer Überblick über verschiedene aktuelle Entwicklungen in der optischen Mikroskopie gegeben und die vorliegende Arbeit in einen Kontext gestellt. Zusätzlich zu den populären superauflösenden Fluoreszenzmethoden werden auch die Perspektiven anderer Techniken und Bildkontraste diskutiert und einige fundamentale und praktische Herausforderungen für die Zukunft betrachtet. Als Nächstes werden in Kapitel 2 die kryogenen und optischen Experimentaufbauten beschrieben, die im Rahmen dieser Arbeit konstruiert wurden. In Kapitel 3 wird dann ein Abriss über die Photophysik organischer Farbstoffmoleküle bei Tieftemperatur präsentiert; es werden auch die in dieser Arbeit verwendeten experimentellen Techniken eingeführt und einige photophysikalische Effekte diskutiert, die bei Tieftemperatur beobachtet wurden, wie etwa Dipol-Dipol-Kopplung zwischen Farbstoffmolekülen. Kapitel 4 gibt dann eine Zusammenfassung sowohl über den mathematischen Hintergrund der Auswertung von Einzelmoleküldaten also auch über die konkrete Implementierung der Algorithmen.

Kapitel 6 stellt die Ergebnisse der superauflösenden Mikroskopie ganzer Zellen bei Tieftemperatur vor. Messungen an gefrorenen CHO-Zellen, die stabil Dopamin-Rezeptoren expremieren, erlauben es den Protomerabstand von Homodimeren dieser Rezeptoren zu bestimmen. In Kapitel 5 wird dann die kryogene Lokalisierungsmikroskopie-methode eingeführt, die durch die wesentliche Verbesserung der molekularen Photostabilität bei Flüssigheliumtemperatur Einzelmoleküllokalisierung auf dem Ångströmniveau erlaubt. Es wird weiterhin die Umsetzbarkeit von Kolokalisierung und kryogener Distanzmessung überprüft durch das Auflösen zweier Fluorophore, die an einer Doppelstrang-DNA in Nanometerabständen angebracht wurden (Kapitel 7). Kapitel 8 präsentiert dann die Ergebnisse zur Auflösung der Positionen mehrerer Fluorophore, die an einem Protein angebracht wurden, mit kryogenischer Lokalisierungsmikroskopie. Durch das Anwenden von Algorithmen entliehen aus der Tieftemperatur-Elektronenmikroskopie kann eine dreidimensionale Dichteverteilung für die Positionen der Fluoreszenzmarker mit einer Auflösung von wenigen Ångström rekonstruiert werden, die eine hervorragende Übereinstimmung mit der erwarteten Kristallstruktur zeigt. Schließlich, in Kapitel 9, werden mögliche technische Verbesserungen, Erweiterungen und Anwendungen der Methode, die in dieser Arbeit vorgestellt wurde, angeregt und diskutiert.

Die in dieser Arbeit vorgestellte Technik erlaubt es Strukturinformation von Molekülen zu gewinnen, die mit den existierenden analytischen Methoden wie Röntgenkristallographie, kryogene Elektronenmikroskopie oder Magnetresonanzspektroskopie nicht zugänglich sind. Diese Ergebnisse stellen einen Auflösungsrekord dar und zeigen, dass optische Auflösung das Beugungslimit um ein Tausendfaches übertreffen kann.

Contents

1 Modern optical microscopy

Light microscopy is one of the oldest scientific tools which is still used in leading-edge research. The diffraction limit in microscopy formulated in the nineteenth century had generations of scientists believe that optical studies of individual molecules and resolving sub-wavelength structures would not be possible. The Nobel Prize in 2014 for super-resolved fluorescence microscopy indicated a clear recognition that these old beliefs have to be reconsidered.

This chapter presents an overview of the recent progress in optical microscopy in order to put the present work into context. Besides the popular super-resolved fluorescence techniques, we discuss the prospects for various other techniques and image contrasts, and we examine some of the fundamental and practical challenges that lie ahead.

The content of this chapter has been published as:
S. Weisenburger and V. Sandoghdar,
Light Microscopy: An ongoing contemporary revolution
Contemporary Physics **56:2**, 123–143 (2015).

Passages of the present text might be nearly identical to the text in the published manuscript.

1.1 Introduction

Optical lenses have been used as magnifying glasses for more than four millennia by the ancient Chinese, Egyptians and Persians [1]. Later, the first compound light microscopes constructed in the 16th and 17th centuries enabled scientists to dive into the microscopic cosmos [2]. In the following 200 years, progress in the fabrication and development of microscopes and lenses came about by empirical optimizations. This changed when the industrialist Carl Zeiss hired the physicist Ernst Abbe who introduced quantitative calculations and theoretical considerations in lens design. In 1873, Abbe published his famous formulation of a fundamental limit for the resolution of an optical imaging system based on the diffraction theory of light [3]. This diffraction limit let generations of scientists believe that optical studies of single molecules and resolving sub-wavelength structures are not feasible. Thus, the focus was put on developing new contrast modalities

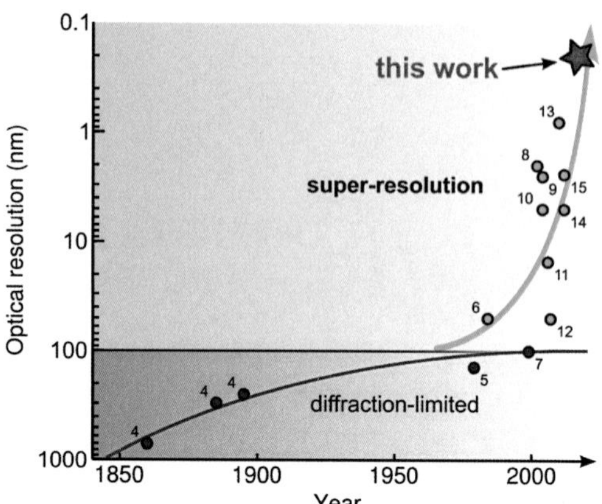

Figure 1.1: Advancement of the optical resolution over time. Data points are taken from Ref. [4–15].

with increased sensitivity to detect small signals and enhanced specificity in order to measure characteristic properties of a specimen.

Finally, during the past four decades, the harsh spell of the diffraction limit was overcome. Several revolutionary methods were conceived and experimentally demonstrated, first by utilizing near-field optical microscopy and subsequently also by far-field methods, which substantially enhance the optical resolution down to the nanometer scale (see Fig. 1.1). The awarding of the 2014 Nobel Prize in Chemistry to Eric Betzig, Stefan Hell and William E. Moerner for their pioneering work in "super-resolved" fluorescence microscopy corroborates its promise for many advanced investigations in physics, chemistry, materials science and life sciences.

Fluorescence microscopy down to the single molecule level has been reviewed in many recent articles and books [16–19]. Despite the immense success of fluorescence microscopy, this technique has several fundamental shortcomings. As a result, many ongoing efforts aim to conceive alternative modes of microscopy based on other contrast mechanisms. Furthermore, having overcome the dogma of the resolution limit, scientists now focus on other important factors such as phototoxicity and compatibility with live imaging, higher speed, multi-scale imaging and correlative microscopy.

1.2 Ingredients for a good microscope

1.2.1 Contrast

Every measurement needs an observable, i.e. a signal. In the case of optical microscopy, one correlates a certain optical signal from the sample with the spatial location of the signal source. Scattering is the fundamental origin of the most common signal or contrast mechanism in imaging. Indeed, when we image a piece of stone with our eyes we

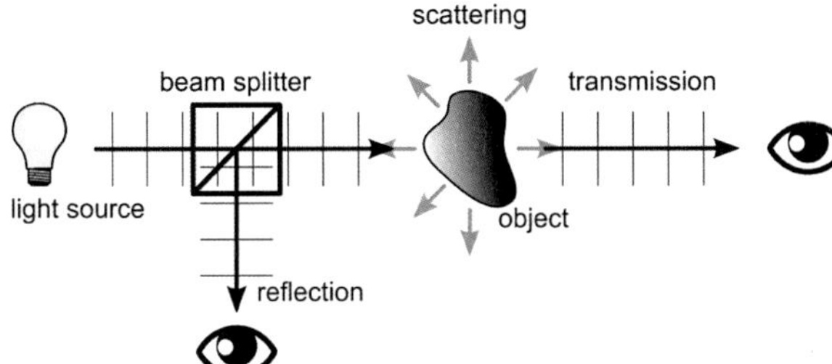

Figure 1.2: Schematic of a basic arrangement for seeing an object in a microscope. The object is illuminated by a light source, and it is observed either in reflection (epi-illumination) via its scattered light or in transmission (trans-illumination) by looking at its shadow.

see the light that is scattered by it, although in common language one might speak of reflection. Scattering also leads to a shadow in transmission (see Fig. 1.2). In conventional microscopy, one speaks of trans-illumination if one detects the light transmitted through the sample and epi-illumination if one detects the signal in reflection.

During the past century, various methods were developed to improve the contrast in standard microscopy. For example, polarization microscopy and phase contrast techniques can be used to examine the anisotropy of birefringent materials such as minerals [20, 21].

Some of the most important contrast mechanisms exploit the spectroscopic information of the sample and thus introduce a certain degree of specificity. The prominent example of these is fluorescence microscopy, where fluorophores of different absorption and emission wavelengths are employed to label various parts of a biological species [22]. Over the years, the developments of fluorescence labeling techniques such as immunofluorescence [23], engineered organic fluorescent molecules [24] and fluorescent proteins [25] have continuously fueled this area of activity. However, fluorescence labeling has many disadvantages such as photobleaching and most importantly the need for labeling itself.

To address some of the limitations of standard fluorescence imaging, scientists have investigated a range of multiphoton fluorescence methods such as two-photon absorption [26]. The strong dependence of multiphoton excitation processes on intensity allows one to excite a small volume of the sample selectively only around the focus of the excitation beam. This leads to minimizing the fluorescence background and photobleaching. Aside from its capacity for optical sectioning, this technique makes it possible to perform tissue imaging because the long-wavelength excitation light penetrates deeper into the biological tissue.

Another effort is to develop microscopy techniques using label-free contrast mechanisms. For example, Raman microscopy generates contrast through the inelastic scattering of light that is selective on the vibrational and rotational modes of the sample molecules [27] and is, thus, very specific to a certain molecule in the specimen. The main difficulty of Raman microscopy lies in its extremely weak cross section. Methods such as coherent anti-Stokes Raman scattering (CARS) [28] or stimulated Raman scattering [29] improve on the sensitivity to some extent although they remain limited well below the single molecule detection level. Some other interesting label-free contrasts are based on the harmonic generation of the illumination or four-wave mixing processes through the nonlinear response of the sample [30]. For instance, it has been shown that collagen can be nicely imaged through second harmonic generation (SHG) [31].

To conclude this section, we return to the most elementary contrast mechanisms, namely transmission and reflection. The fundamental underlying process in these measurements is interference. An early example of employing interference to enhance the the contrast was demonstrated by Zernike in the context of phase contrast microscopy [21]. Here a part of the illumination is phase shifted and then mixed with the light that is transmitted through the sample. Another related technique is known as differential interference contrast microscopy (DIC) put forth by Nomarski [32]. A very similar method to DIC that is somewhat simpler to implement and therefore very common in commercial microscopes is Hoffman modulation contrast [33]. Interferometric scattering contrast was explicitly formulated in the context of detection of nanoscopic particles in our laboratory [34] and has been coined iSCAT [35]. Although simple transmission, reflection, phase contrast and DIC share the same fundamental contrast as that of iSCAT, only the latter has tried to address the issue of sensitivity and its extension to very small nanoparticles and single molecules. This brings us to the general question of sensitivity.

1.2.2 Sensitivity

For any given contrast mechanism, one can ask "how small of an object can one detect?". In other words, what is the sensitivity. To have high sensitivity, one needs a sufficiently large signal from the object of desire, and the trouble is that usually the signal diminishes quickly as the size of the object is reduced. So, one needs to collect the signal efficiently and employ very sensitive detectors. The detector can either be a point detector as it is most often used in scanning confocal techniques or a camera in the case of wide-field imaging. The performance of a light detector can be generally described by its quantum efficiency, the available dynamic range and its time resolution [36].

For the longest time in the history of microscopy the human eye was the only available detector. With its detection threshold of only a few photons it was a better detector than photographic plates and films even long after these became available [37]. Photon counting detectors had emerged by the 1970s, starting with photo-multiplier tubes (PMT) [38] and later followed by semiconductor devices that are able to detect single photons. These single-photon avalanche diodes (SPAD) can nowadays achieve quantum efficiencies above 50 % with timing resolutions on the order of tens of picoseconds. In the 1990s, cameras like the charge-coupled device (CCD) and fast active pixel sensors using complementary metal-oxide-semiconductor (CMOS) technology reached the single-photon sensitivity with high quantum efficiencies. Today, the best CCD cameras using electron multiplication can achieve quantum efficiencies better than 95 % in the visible part of the spectrum with a readout noise of a fraction of a photo-electron.

An additional important issue regarding detection sensitivity concerns the background signal, which can swamp the signal. If the detector dynamic range (the ratio of the highest and the smallest possible amount of light that can be measured) is large enough, one can subtract the background. Temporal fluctuations on the background usually limit this procedure in practice. In general, the image sensitivity can be quantified in terms of the signal-to-noise ratio (SNR),

$$\mathrm{SNR} = \frac{\mu(\mathrm{signal}) - \mu(\mathrm{background})}{\sqrt{\sigma^2(\mathrm{signal}) + \sigma^2(\mathrm{background})}}\,, \tag{1.1}$$

where μ denotes the mean and σ the standard deviation.

The sensitivity of fluorescence microscopy was taken to the single-molecule limit towards the end of the early 1990s [39–41]. The factors leading to the success of this field were 1) access to detectors capable of single-photon counting, 2) the ability to suppress the background by using spectral filters, and 3) preparation of clean samples. Although single-molecule fluorescence microscopy has enabled a series of spectacular studies in biophysics, fluorescence blinking and bleaching as well as low fluorescence quantum yield of most fluorophores pose severe limits on the universal applicability of this technique. The low cross sections of Raman and multiphoton microscopy methods also hinder the sensitivity of these methods. Only in isolated cases, where local field enhancement has been employed, have been reports of single-molecule sensitivity [42,43].

For a long time, single-molecule sensitivity in extinction was also believed not to be within reach because the extinction contrast of a single molecule is of the order of 10^{-6}, making it very challenging to decipher the signal on top of laser intensity fluctuations. Nevertheless, various small objects have been detected and imaged using iSCAT, including very small metallic nanoparticles [34,44], single unlabeled virus particles [35], quantum dots even after photobleaching [45], single molecules [46,47] and even single unlabeled proteins down to a size of 60 kDa [48].

1.2.3 Resolution

One of the most immediate functions that the layperson associates with a microscope is its ability to reveal the small features of a sample. The principle of operation of a microscope is typically described using ray optics. However, when dimensions to be investigated are of the order of the wavelength of visible light, i.e. 400 – 800 nm, we must consider the wave properties of light such as interference and diffraction [3]. Therefore one cannot achieve an arbitrarily high resolution simply by increasing the magnification of the lens arrangement.

Ernst Abbe pioneered a quantitative analysis of the resolution limit of an optical microscope [3]. He considered imaging a diffraction grating under illumination by coherent light. Abbe argued that one would recognize the grating if one could detect at least the first diffraction order (see Fig. 1.3a). In the case of an immersion microscope objective with circular aperture and direct on-axis illumination, the Abbe diffraction limit of resolution d reads

$$d = \frac{\lambda}{\mathrm{NA}} \quad , \tag{1.2}$$

where λ is the wavelength of light and $\mathrm{NA} = n \sin \alpha$ was introduced as the numerical aperture (illustrated in Fig. 1.3b). Here n is the refractive index of the medium, in which the microscope objective is placed, and α denotes the half-angle of the light cone that can enter the microscope objective. Air has a refractive index of about $n = 1$ limiting the NA of dry microscope objectives to less than unity. By filling the space between cover glass and an immersion microscope objective with a high index material, the numerical aperture can be increased.

Already in the original publication, Ernst Abbe discussed how this diffraction-limited resolution can be improved if the illumination comes at an angle with respect to the optical axis, making it possible to collect higher diffraction orders (cf. Fig. 1.3a). In this case, the diffraction limit is determined by the sum of the numerical apertures of the illumination lens and the collection lens. If the angles of incidence and collection are identical, a factor of 2 is obtained, leading to the famous Abbe formula

$$d = \frac{\lambda}{2\mathrm{NA}} \quad . \tag{1.3}$$

Considering that the optical response of an arbitrary object can be Fourier decomposed, Abbe's formula can be used as a general criterion for resolving its spatial features.

At about the same time, Hermann von Helmholtz developed a more elaborate mathematical treatment that he published one year after Abbe [49]. Helmholtz also discussed incoherent illumination and showed that Eq. (1.3) holds in that case too. The intricate details in the difference between coherent and incoherent illumination would require a treatment using optical imaging theory that is beyond the context of this work [50]. In a nutshell, in an incoherent imaging system the sample is being illuminated by an extended light source using a condenser lens, thus one can consider the illumination to be from every direction.

Lord Rayleigh was the first to consider self-luminous objects, which also emit incoherently [51]. Additionally, he discussed different types of objects, different aperture shapes and the similarities in the diffraction limit for microscopes and telescopes. Although Abbe's resolution criterion is more rigorous, a more commonly known formulation of the resolution for spectrometers and imaging instruments is the Rayleigh criterion,

$$d = 1.22 \frac{\lambda}{2\mathrm{NA}} \quad . \tag{1.4}$$

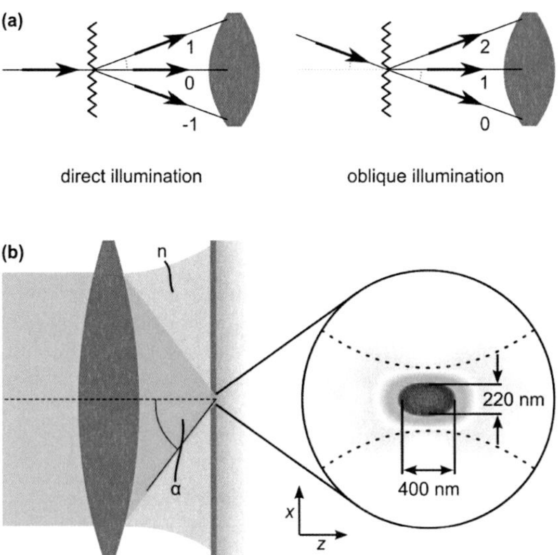

Figure 1.3: Resolution and point-spread function of an optical microscope. (a) Comparison of direct and oblique illumination for Abbe's considerations. In the case of oblique illumination higher diffraction orders can be collected by a lens with the same NA. **(b)** Diffraction-limited point-spread function of an oil immersion microscope objective. Calculated for wavelength $\lambda = 500\,\text{nm}$ and a numerical aperture NA = 1.49. The lateral and axial FWHMs of the PSF amount to about 220 nm and 400 nm, respectively.

It states that two close-lying points are considered resolved if the first intensity maximum of one diffraction pattern of a circular lens described by an Airy disc [52] coincides with the first intensity minimum of the other diffraction pattern [53]. The condition for the validity of Equ. (1.4) is somewhat arbitrary but it comes very close to Abbe's criterion. Rayleigh based his definition upon the physiological property of the human eye, which can only distinguish two points of a certain intensity difference.

In practice, the full-width at half-maximum (FWHM) of the point-spread function (PSF) also provides a useful criterion for the resolution of a microscope because two overlapping PSFs that are much closer than their widths can no longer be resolved. For an immersion oil microscope objective with $\text{NA} = 1.49$ operating at a wavelength of 500 nm, the PSF is about 220 nm (see Fig. 1.3b). The axial width of the PSF is about 2 – 3 times larger, in this case amounting to about 400 nm.

It is worth mentioning that the PSF can also take different forms depending on the employed optical beam. For example, it has been shown that the PSF of a focused doughnut beam with radial polarization can be made somewhat smaller than that of a conventional TEM_{00} mode [54]. The origin of this effect is the vectorial character of optical beams. A thorough discussion of the intricate details in focusing optical beams would go beyond the scope of this work. For an extensive survey of the field of optics, the reader is referred to textbooks like Ref. [55,56].

1.3 Improving the resolution in fluorescence imaging

Fluorescence is one of the most important contrast mechanisms because it offers the possibility of specific labeling. Since the spontaneous emission of a fluorophore does not preserve the coherence of the illumination, the signal is incoherent. As we shall see shortly, one can engineer the illumination to increase the resolution by a factor of two.

1.3.1 Confocal scanning microscopy

One of the possibilities for improving the resolution is confocal microscopy. Although the principle was patented by Marvin Minsky in 1957 [57], it took 20 years until the invention of suitable lasers and the progress in computer-controlled data acquisition opened the door for its widespread use. In contrast to conventional wide-field illumination, where the full field of view is illuminated and imaged onto a camera, scanning confocal microscopy uses spatial filtering to ensure that only a diffraction-limited focal volume is illuminated and that only light from this focal volume can reach the detector. An image is then produced by raster scanning this confocal volume across the sample. Since out-of-focus light is effectively suppressed, the method allows for higher contrast and offers the ability to perform optical sectioning to acquire 3D images.

The lateral size of the PSF can be improved by a factor of $\sqrt{2}$ in confocal microscopy [58]. However, in reality this factor depends on the coherence properties of the imaging light and the finite size of the detection pinhole [58]. The latter is usually set to a value about the size of the point spread function so as to not lose any signal. In practise, the image contrast in confocal imaging is increased causing a better effective resolution [22].

A particularly interesting mode of confocal microscopy is image scanning microscopy, where an image is recorded on a camera at each scan point [59,60]. It has been shown that one can computationally reconstruct an image with a resolution that is improved by up to a factor of two from the resulting image stack.

1.3.2 Structured illumination microscopy (SIM)

The scanning feature of confocal microscopy limits its acquisition speed. An attractive and powerful alternative to improve the lateral resolution of wide-field fluorescence microscopy by up to a factor of two is offered by structured-illumination microscopy (SIM) [7]. Here, the sample is illuminated using a patterned light field, typically sinusoidal stripes produced by the interference of two beams that are split by a diffraction grating.

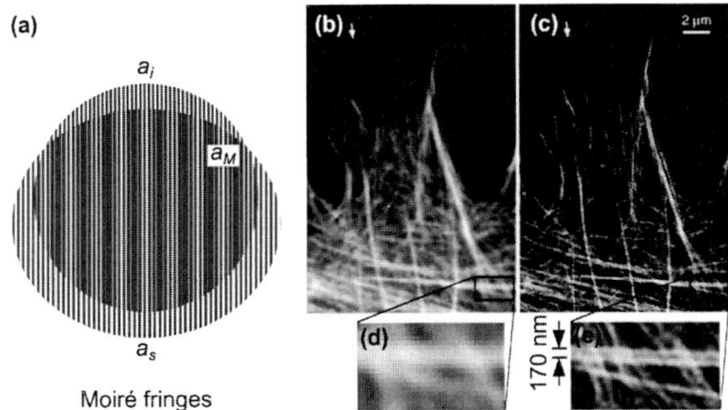

Figure 1.4: Structured illumination microscopy (SIM). (a) Two fine structures (the known illumination pattern and the unknown sample structure) produce a Moirè interference pattern that is imaged by the microscope. The method allows for an improvement of factor 2 in resolution; see text for details. **(b,c)** Actin cytoskeleton at the edge of a HeLa cell imaged by conventional microscopy and SIM, respectively. **(d,e)** Insets show that the widths (FWHM) of the finest protruding fibers (small arrows) are lowered to 110 – 120 nm in **(c)**, compared to 280 – 300 nm in **(b)**. (b-e: Reproduced with permission from [7].)

The resulting image is a product of the illumination pattern and the fluorescence image of the sample. Assuming that the dye concentration follows a certain pattern that can be Fourier decomposed, one obtains a Moirè pattern for each component, resulting from the product of the array of illumination lines and the fluorescence signal (Fig. 1.4a). The key concept in SIM is that the periodicity of a Moirè pattern is lower than the individual arrays.

Let us consider a sample with a periodic array of dyes at distance a_s illuminated by an array of lines spaced by a_i. If we take the angle of the two line arrays to be zero, the period a_M of the Moirè pattern is given by $a_M = (a_s \cdot a_i)/|a_s - a_i|$. Now, consider an objective lens that accepts spatial frequencies k_o. The highest periodicity Moirè pattern that is detected by this objective is $a_M = 1/k_o$, so that the decisive criterion becomes

$$\frac{1}{a_M} = \left| \frac{1}{a_i} - \frac{1}{a_s} \right| \leq k_0 \quad . \tag{1.5}$$

Furthermore, we note that the highest periodicity illumination that is compatible with the objective is $a_i = 1/k_o$. Putting all this information together, one finds that it is possible to detect Fourier components of the sample at $2k_o = 1/a_s$. In other words, one can resolve sample features at a spatial frequency up to twice larger than considered by the Abbe limit.

By recording a series of images for different orientations and phases of the stripe pattern, one can reconstruct the full image with an improved resolution. Interestingly, the resulting Moirè pattern has also a three-dimensional structure, which allows the reconstruction program to obtain an enhanced resolution in the axial direction. Figure 1.4c shows an example of a high resolution image obtained by SIM and the comparison to conventional microscopy (Fig. 1.4b).

There are also approaches that combine structured illumination with other techniques using interference and two opposing objectives to gain high axial resolution ($\mathrm{I}^n\mathrm{M}$, $n = 2, 3, 5$) [61,62]. Recently, SIM has also been combined with light-sheet microscopy (coined lattice light-sheet microscopy) [63]. In an intriguing demonstration, 3D in-vivo imaging of relatively fast dynamical processes is shown using a bound optical lattice as the light sheet. It is worth mentioning that structured illumination strategies cannot yield an extra factor of two improvement for coherent imaging modalities since it yields the same additional information as provided by oblique illumination [64].

1.4 Super-resolution fluorescence microscopy

In 1928, Edward Synge proposed to use a thin opaque metal sheet with sub-wavelength holes to illuminate a sample placed at sub-wavelength separation from it [65]. By scanning the specimen through this point illumination, an image could be recorded with an optical resolution better than the diffraction limit. Technical limitations in the fabrication of nanoscopic apertures, their nano-positioning and sensitive light detection made the experimental realization of a scanning near-field optical microscope (SNOM) only possible in the early 1980s [6].

Near-field microscopy gets around the diffraction limit in a complete and general fashion. The essential point is that the limitations imposed by diffraction do not apply to the distances very close to the source, where non-propagating evanescent fields dominate. These fields contain the high spatial frequency information of the source and sample, but their intensity decays exponentially with a characteristic length of the order of the wavelength of light.

An alternative way of imaging in the near field has emerged in the context of metamaterials. These artificial materials are structures with sub-wavelength-sized unit cells and can be engineered to exhibit intriguing properties such as a negative index of refraction [66,67]. In the year 2000, a so-called perfect lens was proposed by John Pendry using a slab of negative index material [68]. Both propagating and evanescent fields are imaged by this lens yielding a perfect image. Such a perfect lens has not yet been experimentally realized because of the extremely delicate constraints on the properties of the negative index material. There are, however, experimental demonstrations of a superlens in the optical regime where sub-diffraction imaging was shown using metamaterial structures [69]. There have also been efforts for the realization of a hyperlens to project the near field into the far field using a cylindrical [70] or a spherical hyperlens [71]. Fabrication issues, material properties and the requirement that the object of interest must be placed in the near field of the hyperlens, make the practical usage of these interesting imaging techniques very limited although they might possibly find use in nanofabrication and optical data storage.

A young doctoral student in Heidelberg, Stefan Hell, took upon himself not to accept the diffraction limit in far-field microscopy in its usual formulation [72]. In 1991, Hell described an imaging scheme with two opposing microscope objectives with a common focal point [73]. By making sure that the molecules in the common focus are coherently illuminated, the emission can constructively interfere and can be collected from both sides yielding an axial resolution improvement on the order of three to four times. Only a few years later, Hell demonstrated the concept experimentally [74]. In an ideal case, each microscope objective can collect light emitted into a solid angle of 2π, therefore the imaging technique is known as 4Pi microscopy.

1.4.1 STED and STEDish techniques

Realizing that 4Pi microscopy would not provide a means to fundamentally overcome the diffraction barrier, Hell proposed to exploit stimulated emission to deplete (STED) the fluorescence of molecules in the outer part of the illumination and thereby reduce the size of the effective fluorescence spot [75]. The first experimental realization of this idea was reported only a few years later by his group [76]. Hell was awarded the Nobel Prize in Chemistry in 2014 for his achievements in this area.

Upon excitation, a fluorescent molecule is usually brought from its singlet ground state S_0 to a higher vibrational state of the singlet electronic state S_1 (see Fig. 1.5a), which then relaxes on a picosecond time scale to the lowest vibrational level. If the quantum efficiency of the molecule is high, a photon is emitted within a few nanoseconds. Hell had the idea to suppress this fluorescence in the outer part of the excitation beam by stimulating the emission much faster than the nanosecond spontaneous emission in that region. To do this, he used a doughnut-shaped laser beam profile (see Fig. 1.5b) that is overlapped on the excitation focal spot. The stimulated emission takes place at a wavelength that is red-shifted with respect to the main part of the fluorescence line, allowing one to spectrally separate the stimulated emission and the excitation laser line from various fluorescence components.

It is important to note that the depletion doughnut beam itself is also diffraction limited, but the effective size of its hole in the middle can be adjusted by the beam intensity. In other words, the higher the intensity of the depletion beam, the farther one goes into the saturation of the fluorophores. As a result of this nonlinear behavior which is inherent to a quantum mechanical system, the fluorescence PSF can be sculpted. It follows that the resulting sub-diffraction resolution can be described by a modified form of Ernst Abbe's equation after considering the degree of saturation,

$$d = \frac{\lambda}{2\mathrm{NA}\sqrt{1 + I/I_{\mathrm{sat}}}} \quad . \tag{1.6}$$

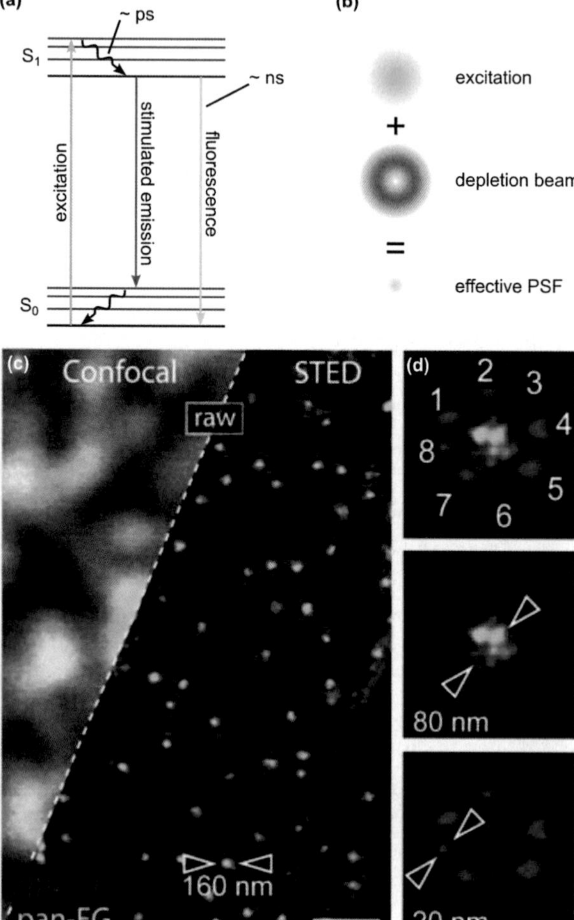

Figure 1.5: Stimulated emission depletion (STED). (a) Overview of the photophysics processes involved: Excitation, stimulated emission and fluorescence. **(b)** Effective sub-diffraction limited PSF as a result of excitation and doughnut-shaped depletion laser beams. **(c)** STED image of immunolabeled subunits in amphibian nuclear pore complex (NPC), raw data smoothed with a Gaussian filter extending over 14 nm in FWHM. Scale bar: 500 nm. **(d)** Individual NPC image showing eight antibody-labeled gp210 homodimers. (c,d: Reproduced with permission from [77].)

Here, I is the peak intensity of the depletion laser and I_{sat} denotes the saturation intensity of the fluorophore. The resolution becomes sub-diffraction limited when the ratio between I and I_{sat} becomes larger than one (see also Fig. 1.5c and d for a comparison of a STED image with a confocal image). It is also important to note that the resolution is in principle not limited. The best resolution that has been demonstrated with STED is about 20 nm in the case of standard fluorescence labeling assays with organic fluorophores [11,77,78] and about 50 nm using genetically expressed fluorescent proteins in live cells [79]. In the case of a very robust emitter such as a nitrogen-vacancy color center in diamond, a resolution down to 2 – 3 nm was successfully demonstrated [15,80].

Sub-diffraction imaging by using a doughnut-shaped excitation pattern for fluorescence suppression can be generalized to other photophysical mechanisms besides stimulated emission. For example, one can exploit the saturated depletion of fluorophores, which can be reversibly switched between a bright state and a dark state. Another possibility is to exploit the saturation of the excited state [81] or a toroidal-shaped decoherence pulse in combination with Femtosecond Stimulated Raman Spectroscopy [82] to achieve a resolution beyond the diffraction limit. The dark state can have a variety of origins such as the ground state of a fluorophore in STED, its triplet state in ground-state depletion (GSD) microscopy [12,83], or a non-fluorescent isomer of an over-expressed fluorescent protein [84]. This general principle was coined "RESOLFT" as an abbreviation for reversible saturable optically linear fluorescence transitions, which is also a pun on the way Germans might pronounce "resolved" [85].

The RESOLFT concept can also be applied to SIM, yielding what is known as saturated SIM (SSIM) [86]. In this regime, a sinusoidal illumination pattern becomes effectively more and more rectangular as the excitation intensity increases. This leads to higher order Fourier terms in the illumination pattern periodicity a_i. Using similar back-of-the-envelope considerations as for standard SIM, $|1/a_i - 1/a_s| \leq k_0$, one can show that spatial frequencies $1/a_s > 2k_0$ become detectable. Indeed, using SSIM, lateral spatial resolutions on the order of 50 nm have been demonstrated [87].

The general class of RESOLFT methods has offered a very clever strategy for circumventing the diffraction limit in fluorescence microscopy. Nevertheless, these methods are accompanied with challenges, which will call for more innovations in the years to come. One of the issues is the fact that so far RESOLFT has been a raster scan technique with a certain temporal resolution, which might limit live cell imaging applications or the study of dynamic processes. This issue can be resolved to some extent by massive parallelization [88–90], e.g. via a square grid similar to structured illumination microscopy.

A second challenge concerns the photophysics of the fluorophores. Excitation to higher electronic states can efficiently compete with the stimulated emission process, opening pathways for photobleaching [91]. The company Abberior has addressed this problem by developing a range of suitable dye molecules and other fluorophores that cover the greater part of the visible spectrum [92]. Recently, other strategies to remedy the issue of photobleaching have been developed such as driving the fluorophores into a second dark state in which it is more resilient to the excitation light [93].

A further area of development will be multicolor applications as it is common in biological fluorescence microscopy. Since already two lasers are necessary for RESOLFT methods, the implementation of two or more color channels is somewhat more complex than in standard fluorescence microscopy. Nevertheless, STED with two different fluorescence labels and two pairs of excitation and STED lasers has been demonstrated [94]. There have also been efforts for sharing the excitation and STED lasers [95], or separating a third fluorophore that lies in the same spectral region by its fluorescence lifetime [96].

Imaging deep in the sample also poses difficulties for RESOLFT microscopy. As the attainable resolution critically depends on the quality of the intensity minimum and phase fronts in the center of the depletion beam, slight changes of the depletion beam profile by scattering deteriorates the imaging performance. This problem can be, in principle, addressed by recent developments using adaptive optics [97,98].

1.4.2 Single molecule localization microscopy

Single molecule microscopy

In parallel to the developments of SNOM in the 1980s and 1990s, scientists worked hard to detect matter at the level of single ions and single molecules. These efforts were originally more motivated by fundamental issues in spectroscopy and to a good part by the community that had invented methods such as spectral hole burning for obtaining high-resolution spectra in the condensed phase [99]. Although already in the 1970s and 80s there had been indications of reaching single-molecule sensitivity in fluorescence detection [100], the first reliable and robust proof came from the work of W. E. Moerner, who showed in an impressive experiment that a single pentacene molecule embedded in an organic crystal could be detected in absorption at liquid helium temperature [101]. Soon after that M. Orrit demonstrated a much better signal-to-noise ratio by recording the fluorescence signal in the same arrangement [39]. The ease of this measurement kick-started the field of single molecule fluorescence detection. However, it was not until 1993 that E. Betzig provided the first images of single molecules at room temperature [40]. Betzig used a fiber-based aperture SNOM to excite conventional dye molecules on a surface. At this point, there was a strong belief that far-field excitation would not be favorable because it would cause a large background. Interestingly, shortly after that R. Zare's lab demonstrated scanning confocal images of single molecules [41]. This achievement was the final step towards a widespread use of single molecule fluorescence microscopy.

A particularly intriguing feature of fluorescence that was revealed by single molecule detection is photoblinking, i.e. the reversible transition between bright and dark states of

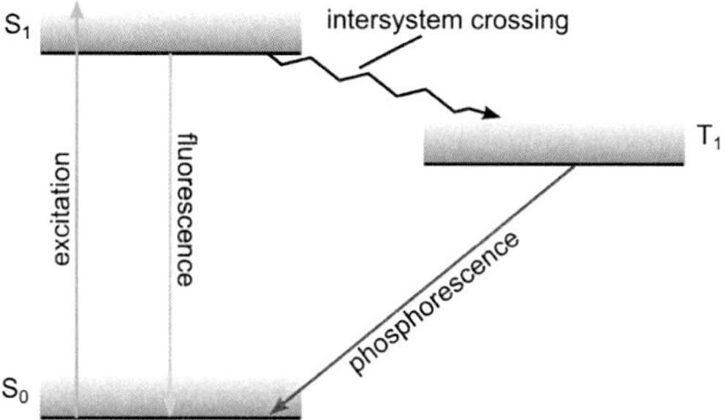

Figure 1.6: Schematic Jablonski diagram illustrating the reason for fluorescence intermittencies in the emission of a single molecule. Intersystem crossing leads to the excitation of a dark, long-lived triplet state.

a fluorescent molecule. Different physical mechanisms may cause fluorescence intermittencies depending on the type of fluorophore as well as its surroundings [102–105]. For example, an excited molecule can undergo a transition to a metastable triplet state with a much longer lifetime than the singlet excited state (see Fig. 1.6). During this time, the molecule is dark because it cannot be excited.

The evidence for triplet state blinking is an off-time distribution that follows an exponential law. However, in some cases the off-time statistics reveal a power law similar to the blinking observed in semiconductor nanocrystals [106, 107]. A proposed mechanism for such a fluorescence intermittency is the formation of a radical dark state [108]. There have been several studies on the topic of blinking, but there is only a limited amount of room-temperature and low-temperature data available and many questions remain open [109–115]. We return to this topic in chapter 3.

Fluorophores also undergo photobleaching, i.e. an irreversible transition to a non-fluorescent product. At room temperature dye molecules typically photobleach within several tens of seconds or a few minutes if sophisticated antifading reagents are used in the buffer. However, the survival times at cryogenic temperatures can go beyond an hour (see chapter 5) or even days in the case of a crystalline matrix. In the special case of terrylene in p-terphenyl a comparable photostability has been achieved even at room temperature [116]. Unfortunately, the combination of aromatic molecules such as terrylene and crystalline host matrices like p-terphenyl is not compatible with the labeling strategies in the life sciences.

Localization microscopy

The idea behind localization microscopy is to find the position of each fluorophore by determining the center of its diffraction-limited PSF with a higher precision than its width. This is accomplished by fitting the distribution of the pixel counts on the camera with a model function that describes this distribution (see Fig. 1.7). The principle

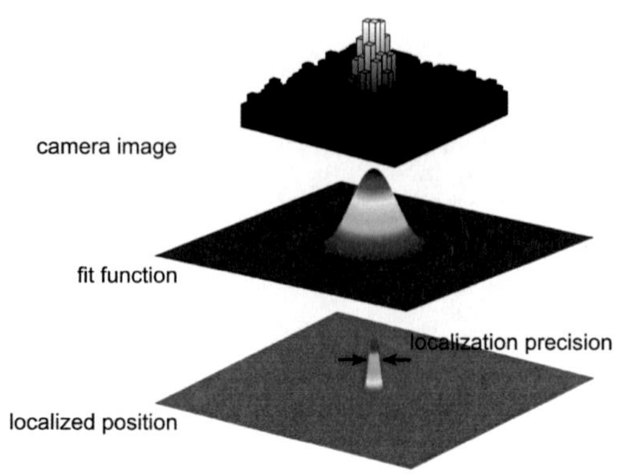

Figure 1.7: Single molecule localization. The position of single fluorophores can be determined by fitting the distribution of pixel counts on the camera with a model function that describes the distribution. The localization precision depends only on the available SNR.

was already conceived by Werner Heisenberg in the 1920s [117] and experimentally demonstrated in the 1980s in the context of localizing a single nanoscale object with nanometer precision [118].

In this scheme, single emitters can be localized with arbitrarily high precision only dependent on the available SNR. The localization precision is mainly determined by the number of photons that reach the detector, the size of the PSF, the level of background noise and the pixel size [119, 120]. The background noise is in turn affected by the luminescence of the cover slip or other materials on the sample as well as the dark counts and readout noise of the camera [36]. The attainable localization precision (σ_{loc}) can be written as

$$\sigma_{\text{loc}} = \sqrt{\frac{s^2 + a^2/12}{N}\left(\frac{16}{9} + \frac{8\pi(s^2 + a^2/12)b^2}{Na^2}\right)} \quad ,$$

using a maximum likelihood estimation procedure with a 2D Gaussian function. This predicts a localization error close to the information limit [121]. Here, N denotes the detected number of photons, s stands for the half-width of the PSF given by the standard deviation of a Gaussian profile, b is the level of background noise and a denotes the pixel size. The limiting factor is typically the finite value of N caused by irreversible photobleaching of the fluorophore. The photon budget of commonly used photoactivatable fluorescent proteins lies in the range of a few hundred detected photons [122], which typically leads to a localization precision on the order of 20 nm. To improve on this limitation, several efforts have optimized the choices of fluorophores and the buffer conditions [13, 123], engineered the dye molecule itself [14], or carefully controlled its environment [124]. The best localization precision for single molecules is reported in this work, just under three Ångströms enabled by cryogenics (see chapter 5).

A particularly powerful tool based on the concept of localization is single particle tracking. Localizing a fluorescent marker or non-fluorescent nano-object of interest as a function of time allows one to study dynamical processes such as diffusion in lipid membranes [125]. Single particle tracking has been performed with various imaging modalities including fluorescence [126, 127], scattering [128, 129] and absorption [130]. However, high temporal and spatial precisions call for a trade-off because smaller integration times and, therefore, lead to a lower signal and lower SNR. Interferometric scattering microscopy (iSCAT) can provide an ideal solution [131], offering up to MHz frame rates in combination with nanometer localization precision in the case of small scatterers like gold nanoparticles.

Interestingly, localization techniques and particle tracking have also found applications in a wide range of studies such as coherent quantum control [132] or cold atoms [133, 134]. For example, identification of atoms down to the single lattice site of an optical lattice has provided an avenue to manipulating single qubits and studying many-body effects like the quantum phase transition from a superfluid to a Mott insulator.

Given that a single molecule can be localized to an arbitrarily high precision, one could also resolve two nearby molecules if only one could address them individually. This was formulated by E. Betzig in 1995 as a general concept [135], but it was already demonstrated experimentally in 1994 by the group of Urs Wild in cryogenic studies [136]. In the latter, the inhomogeneous distribution of the molecular resonance lines allows one to address each molecule separately by tuning the frequency of a narrow-band excitation laser. Our laboratory has recently demonstrated that the same principle of spectral selection can also be used to address single ions in a solid [137]. By combining cryogenic high-resolution spectroscopy and local electric field gradients, it was also shown in our laboratory that two individual molecules could be three-dimensionally resolved with nanometer resolution [8]. In this dissertation we set a new record in three-dimensional optical resolution (see chapter 8).

Several groups tried in the early 2000s different strategies for distinguishing neighboring fluorophores. One example was to use semiconductor nanocrystals with different emission spectra [138] or stepwise bleaching of single molecules [9, 10]. However, extension of these methods to very large number of fluorophores was not practical. The decisive breakthrough came in 2006, when three groups reported very similar strategies based on stochastic photoactivation processes that switched the fluorophores between a dark state and a fluorescent state (see also Fig. 1.8b). Eric Betzig and colleagues called their method photoactivated localization microscopy (PALM) [139], Sam Hess and his team coined the term fluorescence photoactivation localization microscopy (FPALM) [140], and X. Zhuang and her group used the term stochastic image reconstruction microscopy (STORM) [141]. In each case, the fluorophores are placed on the target structure by different labeling techniques. While commonly used antibody-based assays have a label-to-target distance of about 20 nm, nanobodies [142], aptamers [143] or fluorescent proteins [144] can reduce that distance to a few nanometers.

Figure 1.8 illustrates the data acquisition procedure for super-resolution imaging based on single molecule localization. By shining light on the sample with a blue-shifted activation laser beam, one can stochastically switch on a sparse subset of fluorophores. Next, one turns on the excitation laser and collects the fluorescence from the few activated fluorescent molecules until they become deactivated. By adjusting the intensity of the activation beam, one can control the average number of activated fluorescent labels to ensure that PSFs from individual fluorophores do not overlap. One then performs a localization analysis for each recorded PSF to determine the positions of all molecules. The process of activation, recording and localization is then repeated for many other random subsets of fluorophores until one is satisfied with the number of labels for reconstructing a super-resolution image. There are also variations of this acquisition procedure using, for instance, asynchronous activation and deactivation of fluorophores [145] or assays where diffusing fluorophores get activated upon binding to the target structure [146, 147].

In the standard super-resolution imaging modalities such as PALM and STORM, usually photochemistry is employed in order to exert some degree of control on the photoswitching kinetics. This control is necessary since the achievable resolution also depends on the ratio of the on- and off-switching rates of the used fluorophores [148]. Examples of such photochemistry is the chromophore cis-trans isomerization or protonation change in the case of fluorescent proteins [149], and the interplay of reduction and oxidation using enzymatic oxygen-scavenging systems and photochromic blinking for organic dye molecules [150, 151]. The state of the art in resolution for localization microscopy is about 10 nm [14] limited by the total number of photons emitted before photobleaching. However, even sub-nanometer localization precision has been reported in cases where photobleaching could be delayed by using oxygen scavengers [13] or cryogenic conditions reported in the present work (see chapter 7).

Measuring in cryogenic conditions offers the additional advantage of a more rigid sample fixation than chemical fixatives. Indeed, first studies exploring PALM imaging at low temperature have also recently surfaced [152, 153]. The full arsenal of methodologies developed for cryogenic electron microscopy may be applied to prepare samples for cryogenic super-resolution microscopy and even dynamic processes could be studied either by stopping processes at different times or by employing methods like local heating with an infrared laser [154].

A quantitative assessment of the molecules' positions critically depends on both the precision and accuracy of the employed method. Precision in localization microscopy is determined by the standard deviation of the estimated position of an emitter assuming repeated measurements, whereas the accuracy quantifies how close the estimated position lies to the true position. In other words, even if the measurement precision is high, absolute distance information might be compromised by technical sources of bias like pixel response non-uniformity of the camera or sample drift [13].

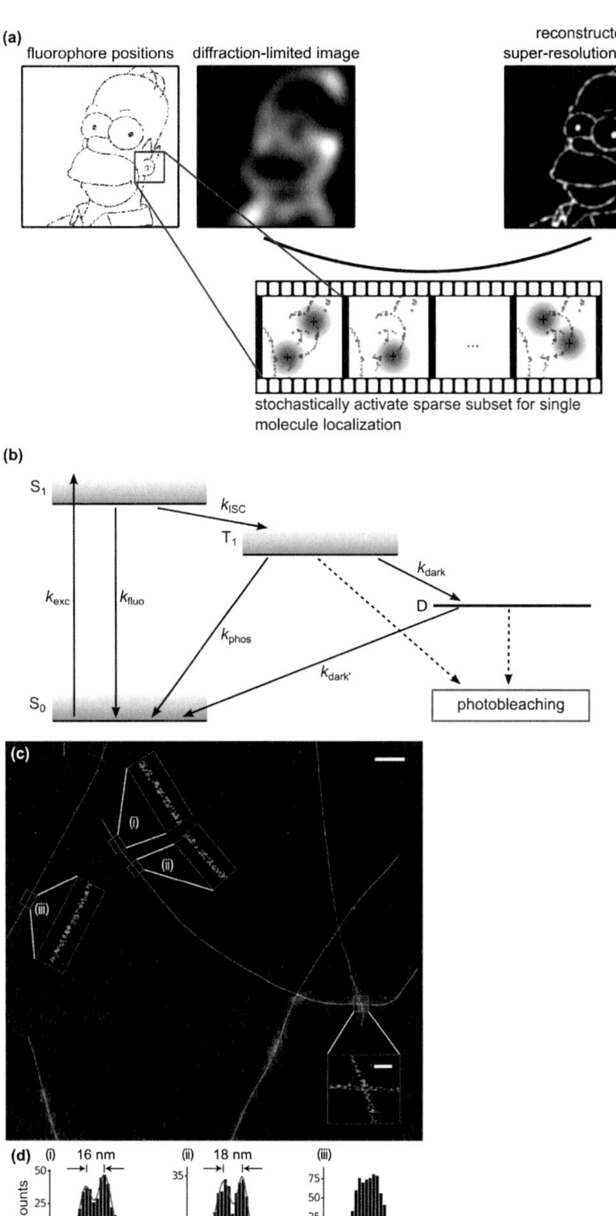

Figure 1.8: Super-resolution imaging based on single molecule localization. **(a)** Illustration of the image acquisition and reconstruction procedure. The target structure is described by the positions of the fluorescence labels. Details in the structure cannot be recognized by diffraction-limited imaging. Repeated stochastic activation, imaging and localization of a sparse subset of fluorophores allows one to reconstruct a super-resolution image. **(b)** Schematic energy diagram for a generic photo-switchable fluorescent molecule as used for example in STORM. **(c)** A STORM image of microtubules (green) with several magnified images shown in the insets. A portion of the corresponding conventional fluorescence image (magenta) is overlaid on the STORM image. **(d)** Segments showing hollow microtubule profiles with inner diameters of 16 – 18 nm. The red curves are nonlinear least-square fits of the distribution to two Gaussian functions. Scale bars: 1 μm, main image; 100 nm, insets. (c,d: Reproduced with permission from [14].)

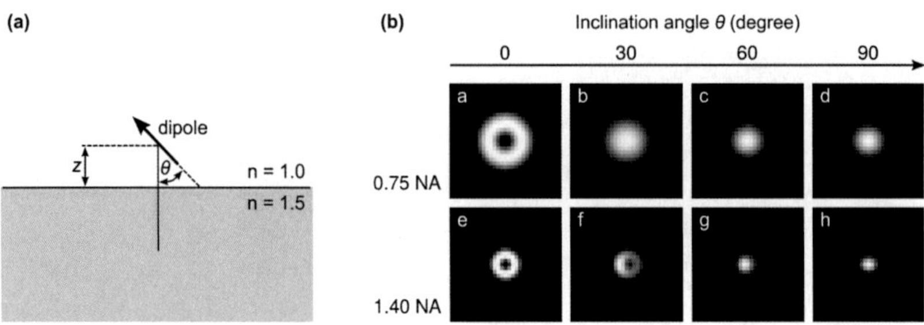

Figure 1.9: Simulations of PSFs of dipoles near an interface. **(a)** Geometry for the simulations. The molecules have an inclination angle θ and the distance between the molecule and the interface is z. **(b)** Examples for PSFs for two different numerical apertures and four inclination angles at z = 2 nm.

An important systematic source of error concerns the dipolar emission characteristics of single molecules. It is known that the image of an oriented dipolar light source deviates from a simple isotropic PSF (see Fig. 1.9). As a result, localization of the individual fluorophores requires a fitting procedure that takes this effect into account [155, 156]. In the presence of nearby interfaces, this can become a nontrivial task (see chapters 4 and 7). In conventional super-resolution microscopy experiments these effects are usually negligible because these systematic errors are much less than 20 nm. The PSF asymmetry is much less pronounced if a microscope objective with low numerical aperture is used [156]. Of course, fitting the data with full theoretical treatment of the PSF or good approximations would also provide accurate values for the position and orientation of the fluorophore [121, 157, 158].

The most severe limit in localization microscopy is the difficulty of high-density labeling. Here it is to be remembered that the image in this method is constructed by joining the centers of the individual fluorophores (see Fig. 1.8). This means that a resolution of 5 nm in deciphering the details of a figure would need at least two fluorophores that are spaced by about 2.5 nm according to the sampling theorem [159]. The first problem with this requirement is the difficulty of labeling at such high densities. Second, once one manages to place the fluorophores at the right place, one faces the problem that such closely-spaced fluorophores undergo resonance energy transfer (homo-FRET) [160, 161]. As a result, the emission cannot be attributed to one or the other fluorophore, and the basic concept of localization microscopy breaks down.

Another issue to be considered is that of accidental overlapping PSFs from neighboring fluorophores. By using appropriate algorithms which can handle an increased fluorophore density faster acquisition times can be achieved [162, 163].

1.5 Ongoing efforts

An important development concerns super-resolution in three dimensions. One possibility to localize a fluorophore along the optical axis is via astigmatism [164]. By inserting a cylindrical lens in the detection path, the PSF becomes elliptical, from the degree of its ellipticity and orientation one can deduce the additional axial position of the fluorophore. Lateral localization precision of about 25 nm with an axial localization precision of about 50 nm was already reported in 2008 [164,165]. Recently, an isotropic localization precision of about 15 nm in all three spatial dimensions was reported using STORM in combination with an Airy-beam PSF [166]. An alternative way to obtain 3D super-resolution is multi-focal plane imaging [167,168]. Here, different focal planes are imaged on various regions of the camera by splitting the fluorescence light and introducing different path lengths. The height can then be deduced from the degree of defocusing. Another approach uses an engineered PSF that encodes the axial position of the emitter in the rotation angle of two lobes, a double-helix PSF [169]. In this work, we use an approach which computationally reconstructs a three-dimensional distribution from a series of lateral projections of randomly oriented samples (see chapter 8), a strategy that is commonly used in single-particle electron microscopy [170].

A crucial requirement for practical biological microscopy is the ability to image many entities simultaneously. Indeed, two-color [171–173] and even three-color super-resolution imaging has been demonstrated [174]. The most convenient and common approach is to label different parts of the specimen with fluorophores of distinct absorption or emission spectra. Considering that the new super-resolution methods, including RESOLFT, PALM, STORM, etc., all rely on the photophysical properties of the fluorophores, it is not a trivial matter to marry them with multicolor imaging. First, fluorescence probes with the desired switching properties must be available with the correct excitation, emission and activation wavelengths. Interestingly, scientists began to develop multicolor super-resolution solutions shortly after the introduction of localization microscopy. A second challenge concerns the crosstalk that is caused by the spectral overlap of the emission bands of different fluorophores, which are typically several tens of nanometers broad at room temperature. A possible solution would be to perform super-resolution microscopy at low temperatures because even though dye molecules suited for labeling in life science do not show lifetime-limited linewidths at cryogenic temperatures, their spectra can become narrower by orders of magnitude (see chapter 3).

As super-resolution optical microscopy becomes a workhorse, it becomes more and more important that it can also handle live cell imaging. In this case, imaging speed and phototoxicity pose important problems. On the one hand, fast imaging often requires a large excitation dose to be able to collect lots of photons in a short time window. High light dose, however, causes the production of free radicals through the photo-induced reaction of the fluorophore with molecular oxygen [175]. Furthermore, excess light also brings about fast photobleaching and short observation times. An interesting approach to minimize these phototoxic effects is light-sheet microscopy [176]. Here, the wide-field detection is disentangled from the illumination, which consist of a thin sheet of light perpendicular to the focal plane. By performing tomographic recordings at different sample orientations, one can then obtain impressive three-dimensional images of whole organs in small animals such as zebrafish [177]. Of course, diffraction limits the thickness of the light sheet to dimensions well above a wavelength, especially if the illumination area is to be large. By employing slowly diffracting beams such as Bessel beams, one can minimize the problem of beam divergence [178]. Light-sheet microscopy is very popular in developmental biology, where super-resolution is less important than large-scale information about the whole system over a longer time. An application example of the technique is the four-dimensional imaging of embryos at single-cell resolution [179, 180].

In the past years, there have been several proof-of-principle demonstrations to obtain fluorescence-free super-resolution images. One possibility is a technique called optical diffraction tomography (ODT) [181, 182]. Here, the sample is illuminated using every possible angle of incidence allowed by the numerical aperture of the microscope objective. Then the intensity, phase and polarization state of the scattered far field are recorded for different angles, and the distribution of the permittivity of the object of interest is reconstructed numerically. Recently, an optical resolution of about one-fourth of the wavelength was experimentally demonstrated [183]. Label-free super-resolution has also been demonstrated using surface-enhanced Raman scattering (SERS) [184], where one performs a stochastic reconstruction analysis on the temporal intensity fluctuations of the SERS signal. Another intriguing method employs ground-state depletion of the charge carriers with a doughnut shaped beam resulting in the transient saturation of the electronic transition by using a pump-probe scheme [185]. There have also been several approaches to achieve super-resolution imaging using the photon statistics of the emitters [186, 187]. Recently, the basic concepts from single molecule localization microscopy also started to be transfered to completely different imaging modalities, such as ultrasound vascular imaging [188].

It is also possible to detect unlabeled biomolecules such as proteins via iSCAT detection of their Rayleigh scattering [48]. In this method the image of a single protein can be localized in the same manner as in Fig. 1.7. Here too, one needs to turn the proteins on and off individually if one wants to resolve them beyond the diffraction limit. In dynamic experiments, the arrival time of each protein can serve as a time tag [48]. The localization precision and therefore the attainable resolution is determined by the signal-to-noise ratio, which in turn depends on the size of the biomolecule in iSCAT. This size-dependent signal also provides a certain level of specificity that in fluorescence modalities can only be achieved by employing different fluorophores.

Another report has used the saturation of scattering in plasmonic nanoparticles. Although the saturation effect in the absorption of small plasmonic nanoparticles has been studied for many years [189], saturation of scattering has only been reported recently [190, 191]. This saturation stems from a depletion of the plasmon resonance. By using the nonlinearity in the saturation and reverse saturation of scattering, it becomes possible to record images with a resolution beyond the diffraction-limit similar to the case of super-resolution imaging in SSIM. By recording images at different light intensities, super-resolved images were obtained and a resolution of $\lambda/8$ was demonstrated [192].

Combination of optical microscopy with other imaging modes such as scanning probe techniques or electron microscopy can offer very useful additional information about the sample. Some of the recent examples of correlative microscopy are the combination of optical super-resolution microscopy with electron microscopy [152, 193, 194] and with atomic force microscopy (AFM) [195–197].

1.6 State-of-the-art resolution

In this chapter we have discussed the recent progress in optical microscopy with a focus on efforts trying to improve the optical resolution. The lateral resolution in the above-mentioned conventional super-resolution methods is currently of the order of 10 – 50 nm, but these techniques are not fundamentally limited by any particular physical phenomenon. The resolution is rather hampered by practical issues, which can be addressed to various degrees in different applications. Thus, only emphasizing a record resolution outside a specific context is not a meaningful exercise.

When performing biological imaging, there are practical implications one has to consider. For example, the amount of laser power that one can shine onto a live cell before it is damaged is orders of magnitude lower than what a diamond sample can take. Moreover, there are important issues concerning labeling techniques and the influence of the label on the functionality of its environment. One subtle point regards the production of free radicals in a photochemical reaction of the excited fluorophore with the surrounding oxygen molecules [198]. To minimize the effect of phototoxicity, it is helpful to acquire images as efficiently as possible. Of course, this is also highly desirable because one gets access to more of the dynamics of the biological and biochemical processes. Recently developed live-cell super-resolution imaging methodologies aim at perturbing the cell as little as possible [63, 199].

If we now relax the requirements for routine biological microscopy, we find that a few experiments have already extended super-resolution microscopy to the nanometer and even sub-nanometer level. The first demonstration of nanometer resolution in all three spatial dimensions used low-temperature fluorescence excitation spectroscopy in combination with recording the position-dependent Stark shift of the molecular transition in an electric field gradient [8]. In another experiment, a distance of about 7 nm was measured between two dye molecules with an accuracy of 0.8 nm by using a feedback loop for the registration of two color channels as well as oxygen-reducing agents [13].

The present work is concerned with achieving Ångström resolution in cryogenic localization microscopy. In Chapter 5 we introduce our cryogenic localization microscopy method which allows for single-molecule localization at the Ångström level due to the substantial improvement of the molecular photostability at liquid helium temperature. Then, we demonstrate the feasibility of co-localization and cryogenic distance measurements by resolving two identical fluorophores attached to a double-stranded DNA at well-defined separations as small as 3 nm (Chapter 7). In Chapter 8 of the present work we will present our results on resolving the positions of multiple fluorophores attached to a protein using cryogenic localization microscopy. By using algorithms borrowed from cryogenic electron microscopy, we can reconstruct a three-dimensional density map for the positions of the fluorescent labels with a resolution of several Ångström.

2 A high-NA cryogenic microscope

This chapter describes the high-NA cryogenic microscope setup that was constructed within the framework of the present work. Some of the experiments that will be described in the following chapters were conducted with the first generation of the cryostat setup, and the later experiments used an improved and more versatile second generation cryostat. The experimental setups evolved during the course of the dissertation work incorporating more functionality for, e.g., confocal beam scanning, spectrally-resolved measurements or time-correlated single photon counting.

2.1 Laser sources

Various fluorophores at different excitation and emission wavelengths were used for the experiments described in this work. Therefore, different laser sources were needed to efficiently excite these fluorophores. For wide-field excitation a continuous-wave (cw) laser beam with sufficient power and reasonable beam quality as well as intensity stability is required. Both diode-pumped solid state (DPSS) lasers as well as some diode lasers meet these requirements.

Table 2.1 gives an overview of the used laser sources with the important properties for this work. While during the beginning of this work the Finesse pump laser (Laser Quantum) was used for green excitation, this laser was replaced with a lower power Coherent Sapphire for the later experiments. For blue and red excitation, the small Toptica iBeam smart diode lasers were used. For measurements that require pulsed excitation, such as fluorescence lifetime measurements, the supercontinuum laser was tuned to the required wavelength. Using the SuperK Varia (NKT Photonics) filter box any part of the visible light spectrum can be cut out with a bandwidth of about 5 nm at reasonable high transmission and about 50 dB out-of-band suppression. The filter box was also used for excitation fluorescence spectroscopy of single molecules using spectral scans.

The required lasers were combined using different methods depending on the experiment at hand. In general, up to four lasers were combined using dichroic mirrors, waveplates and polarization beam splitters, or combinations thereof. The laser light was either coupled into a polarization maintaining fiber or combined into one common free-

Laser	λ (nm)	type	T (ps)	P (mW)	beam quality
Coherent Sapphire	532	OPSL	cw	200	TEM_{00}, $M^2 < 1.1$
Laser Quantum Finesse	532	DPSS	cw	10000	TEM_{00}, $M^2 < 1.1$
Toptica iBeam smart	488	diode	cw	200	TEM_{00}, $M^2 < 1.2$
Toptica iBeam smart	640	diode	cw	150	TEM_{00}, $M^2 < 1.2$
NKT Photonics SuperK Extreme EXR-15	450 - 2400	SC	5 ps @ 80 MHz	1500 (VIS)	TEM_{00}, $M^2 < 1.1$

Table 2.1: Laser sources used for this work. λ – wavelength; T – pulse length; P – power; OPSL – Optically pumped semiconductor laser; SC – super-continuum. Values are taken from the data sheets of the manufacturers [200–203].

space path. While wide-field experiments sometimes required high excitation power that might not be achievable when coupling to a single mode fiber, in confocal experiments usually the mode profile was more important.

2.2 Detectors and sensors

For the main experiment in this work, we usually recorded wide-field imagestacks. For these recordings we used an EM-CCD (iXon+ 897, Andor) camera operated at $-90°$C using thermoelectric (Peltier) cooling and a chiller (FL300, Julabo). We chose this camera mainly because of a large pixel size of $(16\,\mu m)^2$ with unity fill factor and a high quantum efficiency larger than 90 % ($\lambda = 500 - 700$ nm) [204]. For most of the measurements we did not read out the entire chip, and typical frame rates (using frame transfer mode) were usually on the order of a few Hz. Therefore, we used the EM-CCD camera with EM gain switched off. Instead, we used the conventional 1 MHz 14 bit A/D converter to avoid EM excess noise [36] and to reduce readout noise to a minimum. Using this A/D converter the CCD sensitivity is 1.7 e^- per A/D count. The pixel response non-uniformity is < 0.51 %.

For confocal measurements, only a small sample volume was excited and the fluorescence light was imaged onto a single photon counting avalanche diode (PD5C0C, MPD). The timing jitter was determined experimentally to be 68 ps FWHM NIM (Nuclear Instrumentation Module) timing resolution (see Fig. 2.1 for a measurement of the instrument response function (IRF)) and it also features a low dark count rate of about 25 cps.

In order to record emission spectra, we used a grating spectrometer (Shamrock 303i, Andor) comprising of a Czerny-Turner configuration with 300 mm focal length. Our spectrometer was equipped with a 300, 600 and 1,200 lines per millimeter grating

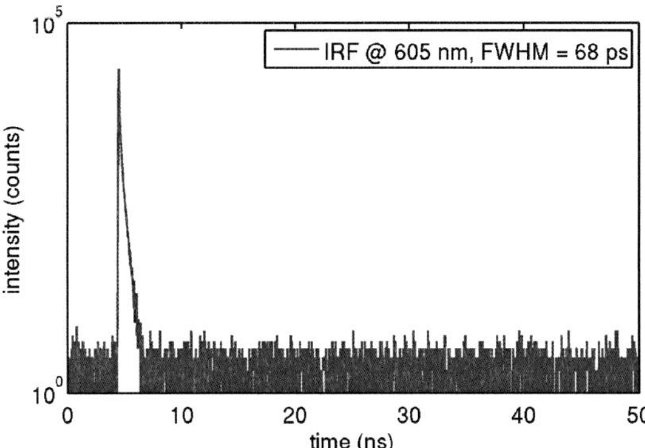

Figure 2.1: Measured instrument response function of the MPD SPAD Recorded at 605 nm using the NIM timing output.

resulting in a wavelength resolution of 0.1 nm. For detection we used an EM-CCD camera (NewtonEM, Andor). The light was guided to the spectrometer by coupling it into a multimode fiber.

2.3 Optical setup

The experimental setup is built on an oscillation damped optical table (TMC CLeanTopII) enclosed on the top and covered on the sides with black PVC walls that can be opened for sample mounting or adjustments. In the middle of the top cover is a filter fan unit that is switched on when no measurements are done to flow clean, dust-free air over the optical table.

A schematic drawing of the optical setup is shown in Fig. 2.2. Depending on the experiment, the required lasers were selected using flip mirrors and they were combined by using the appropriate dichroic mirrors. For lasers that were used for fluorescence excitation, the light also passed through a laser line filter.

The polarization of the excitation light was adjusted to circular by two wave plates. For this purpose, a flip mirror guides the light before entering the telescope to a polarization calibration part of the optical setup. Here, the excitation light intensity is detected by a photodiode after passing through a rotatable polarizer. The two wave plates are adjusted such that the light intensity becomes independent of the polarizer rotation angle.

The excitation light then passes through the wide-field lens which focuses the light through a 4f telescope arrangement into the backfocal plane of the microscope objective. A long-pass filter at low angle of incidence (about 10°) is used to separate excitation and detection. Excess excitation light passing through the beam splitter is removed by a beam dump.

In the first-generation setup a 0.75 NA microscope objective (LD Plan-Neofluar, Zeiss) was used. In the second generation, it was replaced by a 0.9 NA microscope objective (M Plan Apo HR, Mitutoyo). The fluorescence from the sample is collected by the same

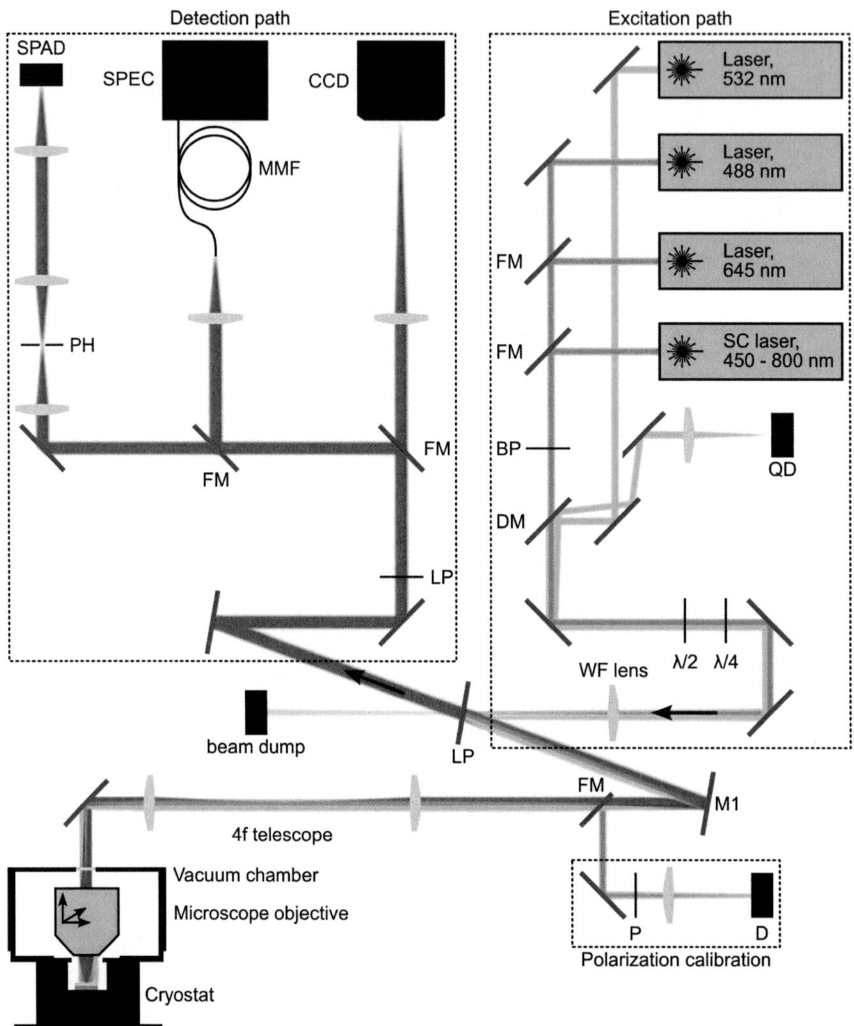

Figure 2.2: Optical path of the experimental setup. SC – super-continuum laser; FM – flip mirror; BP – band-pass filter, DM – dichroic mirror; QD – Quadrature diode; $\lambda/2$ – half-wave plate; $\lambda/4$ – quarter-wave plate; M1 – scan mirror in the confocal configuration; WF – wide-field; LP – long-pass filter; P – polarizer; D – photodiode; MMF – multimode fiber; PH – confocal pinhole.

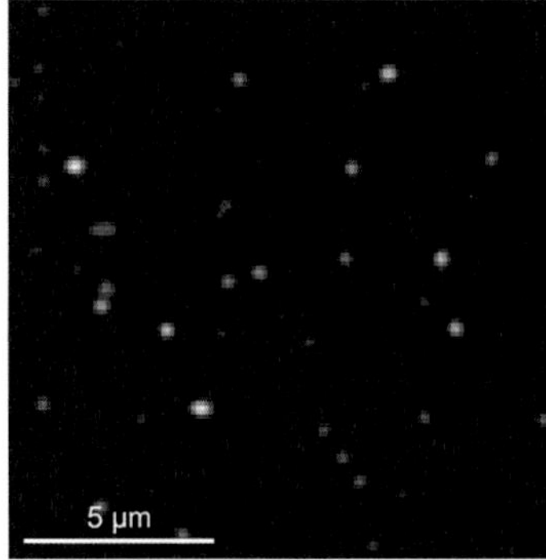

Figure 2.3: Example frame from a wide-field recording. Recording of single streptavidin Atto 488 conjugates adhered to a clean cover slip at 4 K.

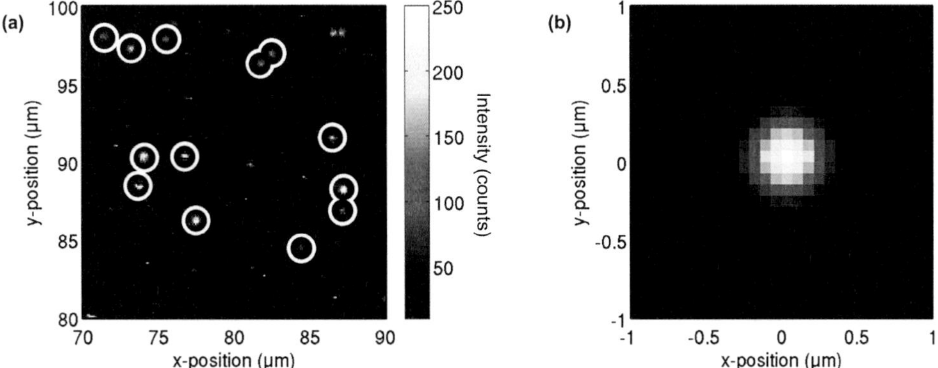

Figure 2.4: Example of a confocal scan. Confocal scan of single Alexa 532 molecules **(a)** and scanned image of a single 200 nm bead **(b)**. White circles mark molecules that were selected for further measurements.

microscope objective in epi-mode. After passing through the beam splitter, the remaining excitation light is filtered by another long-pass filter before the fluorescence is sent to a detector. In the normal operation mode, the light is send to a CCD camera which is placed as close as possible to the cryostat. Figure 2.3 shows an example frame from such a wide-field recording. The figure shows single streptavidin proteins labeled with Atto 488 adhered to a clean cover slip at 4 K.

In confocal mode, light was guided to the beam splitter with a single-mode, polarization-maintaining fiber to ensure a high-quality diffraction-limited spot. By means of a flip mirror, the collected light can be focused through a confocal pinhole before it is detected by a SPAD. Figure 2.4a shows an example of a confocal scan of single Alexa 532 molecules adhered to a clean cover slip at 4 K. Each bright spot represents a single fluorescent molecule. The image was obtained by raster scanning a confocal laser beam across the sample and recording the fluorescence signal as function of the excitation

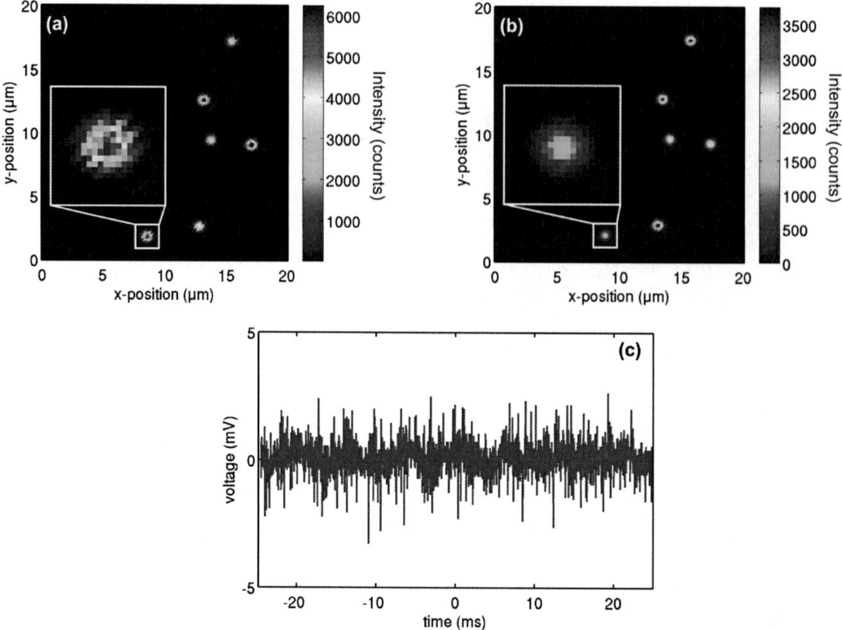

Figure 2.5: Influence of noise on the steering mirror voltage on the scanned images. Confocal image of 200 nm fluorescent beads before **(a)** and after **(b)** the measures to reduce 50 Hz noise on the steering mirror voltage. Insets show a zoom on the PSF of one fluorescent bead. **(c)** 50 Hz noise after reduction measures.

laser position. Pixel dwell times were on the order of milliseconds. This means that the image is not recorded at one point in time and reflects the temporal dynamics of the fluorophores. Indeed, we can observe fluorescence intermittency in the diffraction-limited spots. The excitation power for confocal scans was usually on the order of a few μW in a diffraction-limited spot. Panel (b) in Fig. 2.4 shows a much finer scan over a fluorescent 200 nm bead.

The laser beam scanning was realized by replacing the mirror M1 before the 4f telescope by a piezo scan mirror (S-334, Physik Instrumente). It turned out that electrical noise on the input voltages for the piezo steering mirror disturbed the confocal scans in the beginning. A noise on the order of 10 mV corresponds to a position noise of about 100 nm leading to a visible distortion in the scanned image (see Fig. 2.5a). The frequency of this noise was determined to be 50 Hz, revealing the influence of the power line. By introducing a galvanic isolation of the detection electronics, using a 3:1 voltage divider for the input voltage, and keeping the connector cable length at a minimum to avoid in-coupling, we could reduce the noise to below 2 mV rms (Fig. 2.5c). This measure reduced the noise on the steering mirror voltage inputs to a level that is no longer visible

Optical element	First-generation setup, detection efficiency (%)	Second-generation setup, detection efficiency (%)
Objective, collection efficiency	16	29
Objective, transmittance	80	80
4x Ag-mirror, transmittance	90.4	90.4
3x VIS lens, transmittance	98.8	98.8
2x long pass, transmittance	94	94
CCD, quantum efficiency	95	95
Complete system	**10**	**18**

Table 2.2: Estimation of the detection efficiencies. The detection efficiencies are estimated by considering the transmittance and reflectance values of the individual elements. The values are taken from the data sheets of the manufacturers [204–208].

on the confocal scans (see Fig. 2.5b).

For the measurements with the second generation setup (see below), we also implemented a focus lock using an extra laser that is shifted with respect to the optical axis. Therefore the reflection angle from the cover slip is a function of the distance between the cover slip and the microscope objective. By detecting the positional shift of the reflected beam with a quadrature diode (QD) and introducing a feedback loop, the focal position can be locked.

By having a look on the individual transmittance and reflectance values of the elements in the detection path, we can estimate the detection efficiency, i.e. the probability of detecting a photon that was emitted in the focus of the microscope objective. Table 2.2 gives the efficiencies of the respective elements: The collection efficiency can be computed as the integrated intensity emitted by a dipole [209] into the solid angle given by the numerical aperture of the objective divided by the integrated intensity emitted into the full solid angle. The transmittance values of the microscope objectives is about 80 % in the green to red region of the visible spectrum in both cases [205,206]. The reflectance of one interface of a lens is below 0.4 %. The reflectance values of the protected silver mirrors and the long pass filters are taken from the manufacturer data sheet. The product of the individual efficiencies yields about 10 % and 18 % as detection efficiencies of the complete systems, respectively.

Figure 2.6 depicts the evolution of the experimental setup between the years 2011 and 2015. During the course of the present work, almost every part of the experimental setup was improved or replaced, and new features were added. Panels (a-d) show the evolution of the first generation cryostat setup whereas in panel (e) the second generation setup is shown. In the following section, we will have a closer look at these two cryogenic setups.

Figure 2.6: Evolution of the experimental setup. Photographs of the experimental setup from 2011 until 2015.

2.4 Cryogenic setup

2.4.1 First-generation cryostat setup

The cryogenic experimental setup must fulfill a series of requirements on mechanical stability, geometry and compatibility with the optical setup. Since we expect to conduct experiments with acquisition times ranging from minutes to hours, lateral and vertical sample drift as well as mechanical vibrations should be minimized. All homebuilt parts were optimized in that sense. Furthermore, there are some restrictions caused by the cryostat window or the sample size on properties of the used microscope objective such as working distance or cover glass correction. Ideally, the microscope objective should have the highest numerical aperture that the system allows for.

Figure 2.7 shows a schematic drawing of the cryostat and microscope objective arrangement of the first generation cryostat setup. The used continuous flow cryostat (ST-500, Janis) is mounted on the optical table and cooled to 4 K using liquid helium. The sample is a cover slip with the molecule of interest spin coated on top of it either embedded in a thin (100 nm, verified by atomic force microscopy) polymer layer or directly adhered to the interface. This cover slip is fixed to the cold finger using low temperature grease (Apiezon N). A 0.5 mm thick glass window at a distance of 0.6 mm from the cover slip allows for optical access. We used a long-working distance microscope objective (LD Plan Neofluar, Zeiss) that features a correction collar for cover glass thickness between 0 – 1.5 mm for BK7 glass, keeping in mind that the difference in the refractive index between fused silica glass ($n = 1.46$) and BK7 glass ($n = 1.51$) must be taken into account.

The sample was in vacuum and cooled by thermal contact to the cold finger during measurements. By means of a transfer system, liquid helium was flown through the cryostat from an external helium dewar. The helium flow rate was minimized such that the temperature of 4 K could just be held by the system to reduce vibrations. The cryostat features a heating system including feedback loop which also allows to stabilize the temperature to a given set point.

In the first experiments, the microscope objective was mounted in a homebuilt construction that allows for nano-positioning in all three dimensions (see Fig. 2.8a – c for a photograph and schematic drawings). Part of the setup was built by a master student who worked on the project and is described in more detail in Reference [210]. Briefly, the microscope objective is mounted in a piezo focusing system (Physik Instrumente) which is connected to an xy-piezo system (NPS-XY-100A, Queensgate). The xy-piezo system was replaced by a solid aluminum part in the early stage of this work since it was usually not used. For coarse focusing, the whole construction could be raised by a focusing ring that is running on an outer thread. During normal operation, the coarse position was usually fixed by fixation screws. The setup stood on three pins which allow for axial adjustment of the microscope objective with respect to the optical axis.

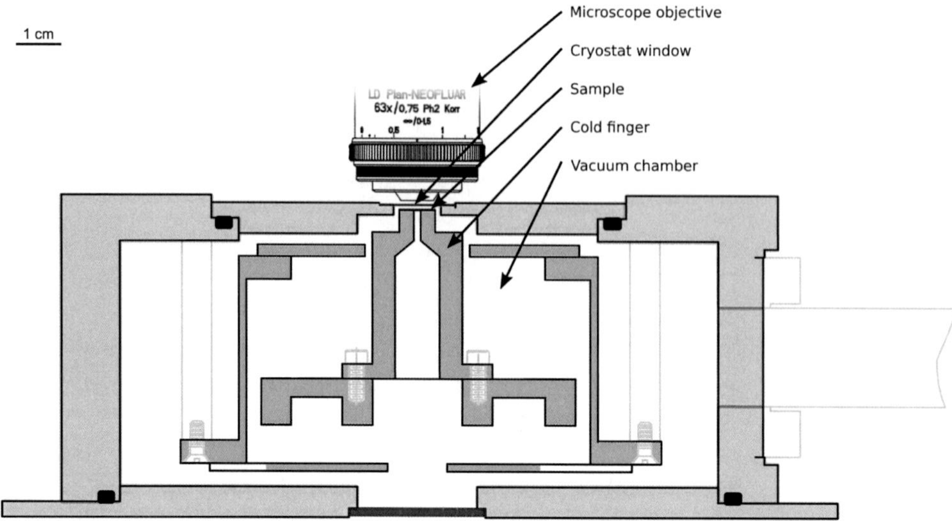

Figure 2.7: Schematic drawing of the cryostat and microscope objective arrangement. Picture of the microscope objective taken from Zeiss [205].

The mechanical stability of the experimental setup was determined by tracking of fluorescent beads (diameter 200 nm) at a frame rate of 750 Hz. Figure 2.8d shows a typical drift trajectory that was obtained over 1 h. By computing the fast Fourier transform (FFT) of the x-component and y-component of this trajectory, we can obtain a mechanical vibration spectrum (see Fig. 2.8e) which confirms that the amplitudes of mechanical vibrations are on the order of one nanometer. It should be pointed out that vibrations on a much faster time scale than the measurement frame rate only cause a slight broadening of the point spread function and no bias on the position.

2.4.2 Second-generation cryostat setup

During the course of this dissertation, a second generation cryostat was designed and constructed. The new experimental setup can be used with a microscope objective of higher NA because it is mounted inside the vacuum chamber to bypass the cryostat window. Furthermore, the new cryostat setup was designed to be compatible with the usage of a solid immersion lens (which will be used in future experiments). It also allows for a higher degree of experiment automation and is more flexible in terms of space to incorporate additional functions in the experiment like prism-based total internal reflection illumination. For this purpose we also constructed a new cold finger made from gold-plated oxygen-free high thermal conductivity copper, and raised the radiation shield to minimize thermal radiation coupling. We furthermore placed a temperature sensor on top of the cold finger to confirm that the cryostat can still reach a temperature

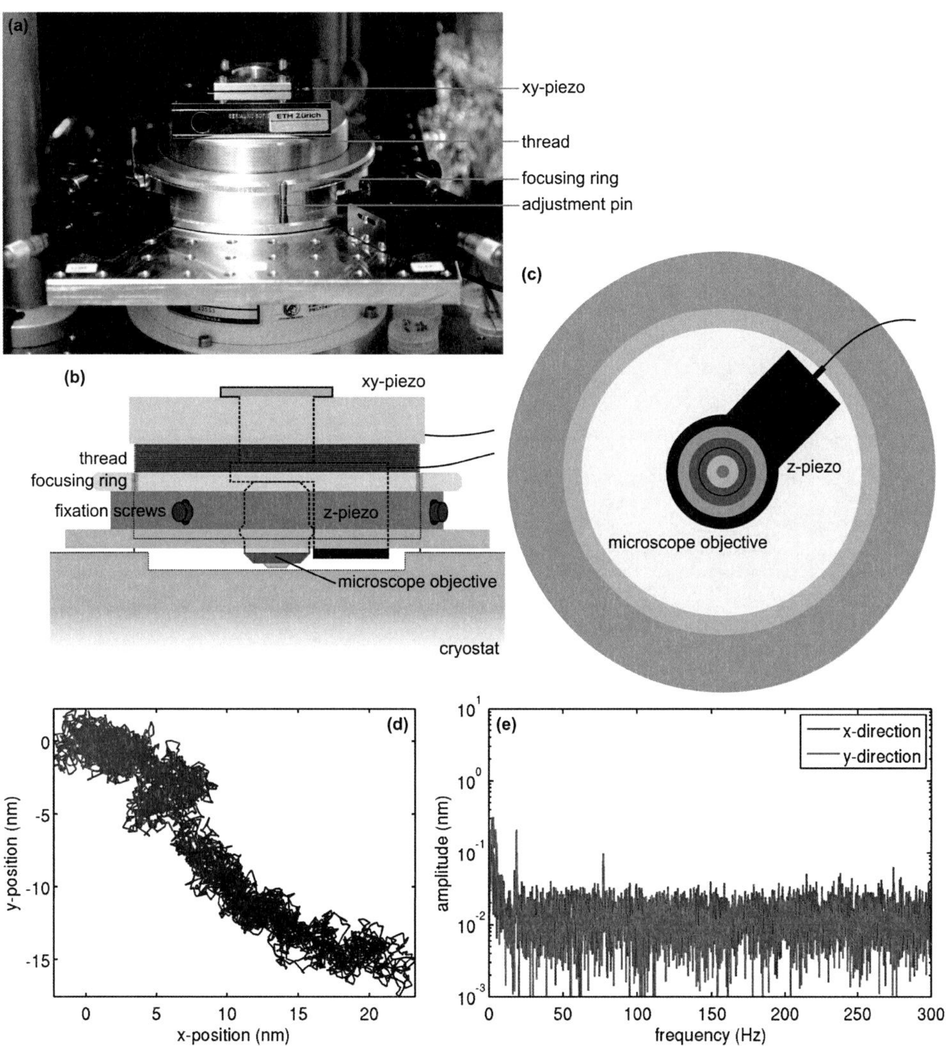

Figure 2.8: First-generation cryostat setup. Photograph **(a)**, side-view schematic drawing **(b)** and top-view drawing **(c)** of the microscope objective mount. The mount stands on three pins that can be adjusted for axial alignment of the microscope objective with respect to the optical axis. (Drawings from Reference [210].) **(d)** Drift trajectory obtained over 1 h (the gradient from blue to black encodes time) and **(e)** mechanical vibration spectrum of the first generation cryostat setup.

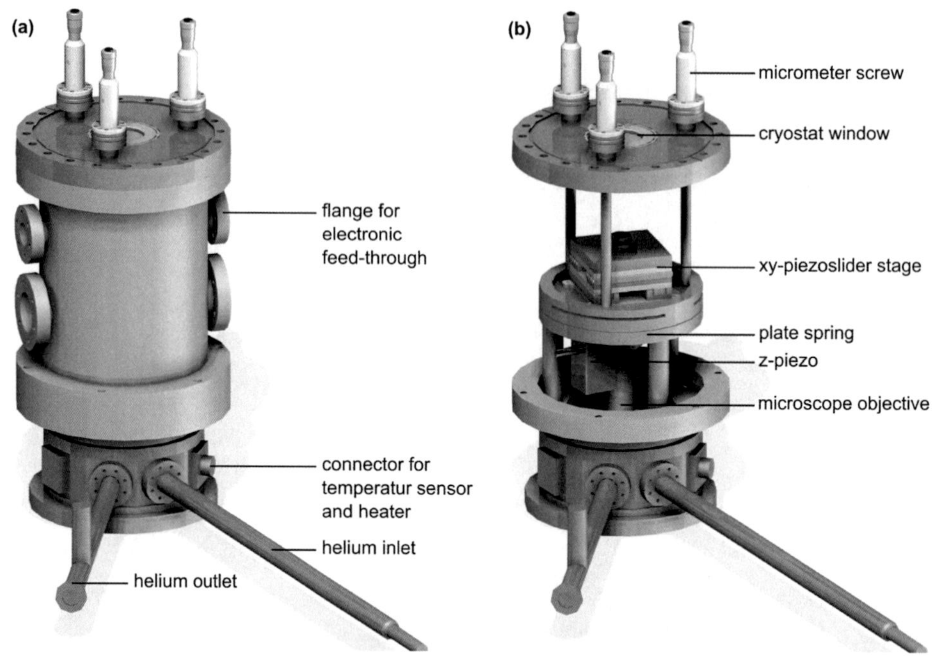

Figure 2.9: Render of the MK4-assembly. Assembly with **(a)** and without vacuum chamber **(b)**.

below 6 K.

Figure 2.9 shows a render of the final (MK-4) assembly of the second generation cryostat. A vacuum chamber was engineered to sit on the basis of the same cryostat system (ST-500, Janis) as before. Since in this configuration there is no window between the sample and the microscope objective, it does not require a correction ring anymore. We can therefore use a microscope objective with higher NA of 0.9 (M Plan Apo HR, Mitutoyo). It is mounted on a piezo-focusing system (Piezosysteme Jena) which, in turn, is mounted on a custom-built 2D piezo-slider system (Smaract). The latter allows selecting different fields of view as well as the alignment of the microscope objective on the optical axis. The 2D stage is attached to a platform that can be moved by means of micrometer screws for coarse adjustment of the focus. Note that the cover slip must be placed face up in this configuration, thus it is likely that the actual temperature on the interface is higher than specified.

In the first versions (MK-1 and MK-2) of the cryostat extension, we used a system of stainless steel / brass guidance rods for coarse adjustment of the focus, which proved to have insufficient mechanical stability. Furthermore, only two micrometer screws were used which axial motion was translated onto the three-pin arrangement. This construction showed a high stability in one axis but a considerably lower stability in the

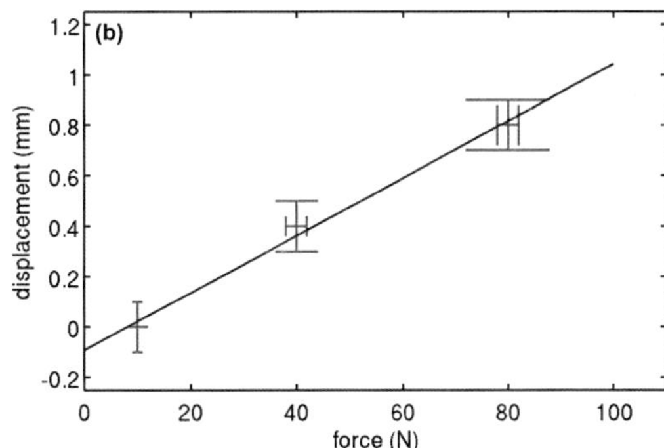

Figure 2.10: Plate spring used for the MK-3+ models. **(a)** Picture of the plate spring after machining the slits to adjust the force curve. **(b)** Calculated (blue) and measured (red) force curve of the machined plate spring. The force curve was designed for maximal travel range of the plate spring, taking into account the maximal force that the micrometer screws can provide.

other direction. In the final (MK-4) version of the setup, three micrometer screws work against a plate spring (Christian Bauer GmbH) (see Fig. 2.10a for a photograph) with custom designed force curve (c.f. Fig 2.10b) obtained by machining slits into the plate spring. This configuration strongly increased the mechanical stability.

The mechanical drift and vibration was measured as before. Figure 2.11 shows the result for the final (MK-4) version of the second generation experimental setup. Again, our measurements confirm that the amplitudes of the mechanical vibrations are on the order of one nanometer. As discussed above, vibrations on a much faster time scale than the measurement frame rate only cause a slight broadening of the point spread function still allowing for Ångström localization precision.

2.5 Experiment control and data acquisition

The experiment control and data acquisition (DAQ) was mainly performed with three computers (see Fig. 2.12). Experiments are in general controlled, processed and triggered by central LabVIEW (National Instruments) computer programs running either on PC 1 or PC 2. In the case of the main experiment wide-field image stacks are recorded. Here, PC 1 reads the CCD camera via its proprietary PCI-Express card. A central LabVIEW program controls the data acquisition via a DAQ card (PCI-6229, National Instruments) and two connector interfaces (BNC-2110, National Instruments), as well as through the other input/output interfaces of the computer. The central program controls the shutters, and has an implemented feedback loop for focus locking using the input signal from the quadrature diode and feeding the z-piezo. The quadrature diode can also be read by an oscilloscope (Tektronic 60 MHz) for alignment purposes. It also has the option to monitor

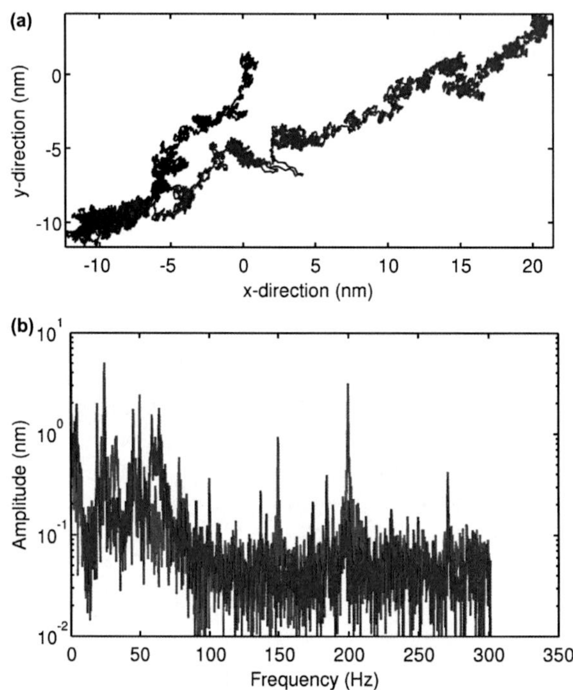

Figure 2.11: Mechanical stability of the MK-4 second generation cryostat setup. **(a)** 1 h drift (the gradient from black to blue encodes time) and mechanical vibration spectrum **(b)** of the MK-4 model of the second generation cryostat.

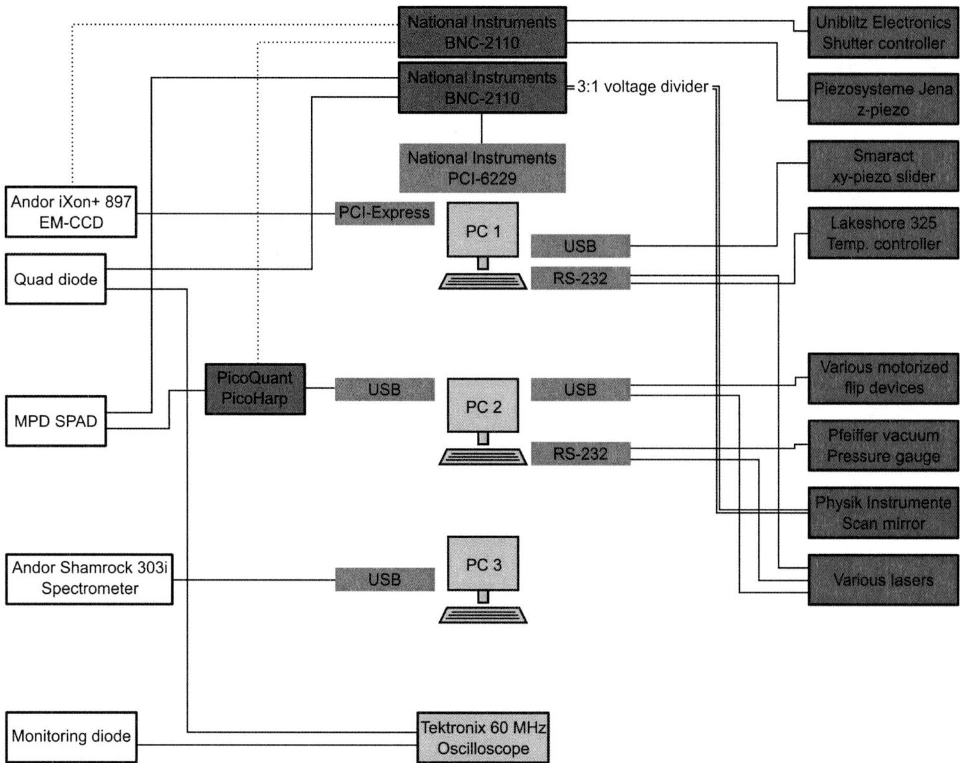

Figure 2.12: Schematic overview of the experiment control and data acquisition. Two computers (PC 1 and PC 2) record data from acquisition devices (white) via external devices (gray) or computer I/O (red) and feed controller devices (blue). Data connections are visualized by solid lines, trigger signals by dotted lines.

the cryostat temperature and automated acquisition of multiple field of views by moving the field of view using the xy-piezo slider. For this experiment, the internal trigger of the camera is used for experiment synchronization.

For confocal measurements, again, a central LabVIEW program controls the data acquisition. The signal from the SPAD can either be recorded through the TTL output which is connected to the DAQ card of PC 1. The LabVIEW program was used to record confocal scans by changing the input voltage of the piezo scan mirror. For the actual time-correlated single-photon counting (TCSPC) measurements such as fluorescence lifetime or autocorrelation measurements, a TCSPC unit (PicoHarp, PicoQuant) recorded the NIM output of the SPAD and it was connected to a second computer (PC 2). In this scheme, PC 1 allowed to select a molecule in the recorded confocal scan and then trigger the TCSPC measurement on PC 2.

Furthermore, PC 2 was also used to switch on and change settings of the various lasers of the experimental setup and to remote control various motorized flip mirrors and filter wheels. The spectrometer (Shamrock 303i, Andor) is controlled by a separate computer (PC 3) using Andor Solis.

3 Low temperature photophysics of organic dye molecules

The success of single molecule imaging modalities and in particular super-resolution fluorescence microscopy critically depends on the photophysical properties of the used fluorescent probes. Recently, cryogenic single-molecule methods received increasing attention in various fields of biological and biophysical research. In the case of super-resolution imaging, they offer superior structural preservation during sample fixation, compatibility for correlative imaging with cryogenic electron or X-ray microscopy, and the promise for Ångström optical resolution. However, there has been only a limited number of low temperature studies on photoblinking and photobleaching or organic dye molecules that are compatible with labeling in life sciences.

In this chapter, we report on photophysical properties at cryogenic temperatures of various organic dye molecules that are commonly used for labeling of biological samples. We present fluorescence excitation and emission spectra, and we show that these dye molecules can have considerably narrower linewidths at T = 4 K compared to room temperature. We investigate their photoblinking behavior and discuss effects such as fluorescence quenching and changes in the temporal evolution of the fluorescence lifetime in molecular complexes labeled with multiple fluorophores.

The content of this chapter is part of the following manuscript:
S. Weisenburger and V. Sandoghdar,
Low temperature photophysics of organic dye molecules
in preparation.

Passages of the present text might be nearly identical to the text in the manuscript to be submitted.

3.1 Introduction

As we have discussed already in chapter 1, fluorescence microscopy has tremendously impacted research in various fields and, in particular, advanced our knowledge and understanding of structures and processes in life sciences [22, 25, 151, 211]. The use of small organic dye molecules as fluorophores turned out to be a key factor that enabled this progress. Organic dye molecules offer high brightness, minimal perturbation of the biological system due to their small size, and are compatible with a multitude of established labeling approaches [212]. They greatly facilitated single molecule studies which allow us to overcome the limitations of ensemble averaging, making information available that would otherwise stay obscured in bulk experiments [213–215]. Single molecule methods have proven to be especially useful in quantitative studies and in investigating dynamic and transient processes. Another important application of single molecules are the localization-based super-resolution methods [139–141].

These super-resolution techniques have a Janus-faced relationship with the used fluorescent labels. On the one hand, they critically rely on the molecules showing a stochastic on/off behavior in their emitted fluorescence intensity called photoblinking. This fluorescence intermittency allows to sequentially image and localize multiple fluorophores which are located within a diffraction-limited spot (see chapter 1). On the other hand, the localization precision is inversely proportional to the available signal-to-noise ratio which is limited by irreversible photobleaching of the fluorophores at room temperature [119]. The limited amount of light, which can be recorded from each fluorophore, limits the localization precision and thus the achievable resolution to values on the order of ten nanometers [14].

Inspired by the success of room-temperature super-resolution microscopy and the contributions of cryogenic electron microscopy [216], our laboratory has pursued cryogenic single-molecule experiments, where the number of collected photons before photobleaching can be increased by orders of magnitude (see chapter 5). In this dissertation, we will demonstrate single-molecule localization at the Ångström level (chapter 5) and nanometer distance measurements at cryogenic temperatures by resolving two fluorophores on the backbone of a double-stranded DNA (chapter 7). There have also been first reports on super-resolution imaging using the PALM modality at cryogenic temperatures [153] and also in correlation with electron microscopy [152]. Although it has been known since the dawn of single-molecule spectroscopy that cryogenic temperatures enhance the molecular photostability of fluorophores [39, 101], there are still many open questions, especially concerning bio-compatible fluorophores. Moreover, the low temperature photophysics of fluorescence intermittency and photobleaching of these fluorophores are largely unknown. A deeper understanding of the interplay of these effects would allow for the optimization of cryogenic super-resolution schemes.

In this chapter, we report data on photophysical effects of a selection of rhodamine, cyanine and carbopyronine derivatives recorded at liquid helium temperature. We investigated the statistics of fluorescence intermittency by computing on/off-time histograms and second-order autocorrelations from time-correlated single photon counting (TCSPC) data. We furthermore measured the temporal evolution of the fluorescence lifetime of molecular complexes labeled with multiple fluorophores. We observed a quenching effect which is dependent on temperature and fluorophore separation. We also recorded fluorescence emission spectra and fluorescence excitation spectra of single molecules at cryogenic temperatures and observed a strong heterogeneity in the linewidths.

3.2 Materials and methods

3.2.1 Experimental setup

Experiments were performed at cryogenic or room temperature using a home-built epi-fluorescence or confocal microscope (described in chapter 2). For saturation measurements, the power could be controlled by a motorized gradient neutral density filter wheel. The fluorescence was collected in reflection and separated from the excitation light by a glass wedge (Thorlabs) at low angle of incidence and an appropriate long-pass filter (RazorEdge, Semrock or 3rd Millenium). The excitation light that is being transmitted through the glass wedge is sent to a diode and used either for monitoring the laser power stability or as a reference signal.

3.2.2 Sample preparation

Samples were prepared by spin-coating (10 s at 1000 rpm followed by 30 s at 3000 rpm) a dilute solution of the dye molecule or dye molecule complex of interest (500 nM) with Tris-EDTA buffer (Fluka, BioUltra (10 mM Tris-HCl; 1 mM EDTA; pH 7.4)) in polyvinyl alcohol (PVA, Aldrich (M_w 9,000 – 10,000, 80 % hydrolyzed)) on fused silica cover slips (thickness 0.17 mm, Esco Products). The substrates were cut to about $7 \times 7\,\mathrm{mm}^2$ square pieces before they were thoroughly cleaned by alternating oxygen plasma and rinsing with deionized water as well as non-halogenated solvents (acetone, ethanol, methanol, and 2-propanol, in that order). The samples were placed in the cryostat chamber immediately after preparation.

We measured samples with three different Alexa Fluor dyes (Invitrogen) and three different Atto dyes (ATTO-TEC), as well as Alexa Fluor dyes attached to streptavidin proteins (Invitrogen) (see also table 3.1 for an overview). For the DNA rulers, a DNA strand with the sequence GCGAGTTCCACCTACCCTGCCTAAGCCTGTATC(C6dT)GTCA was labeled at position 34, where C6dT represents the modified thymidine deoxynucleosides with a flexible linker containing six methylene groups and a terminal amine group. This strand was annealed to different labeled oligonucleotides with each construct resulting in a different sequence separation. The sequence of the complementary second strand was CGCTCAAGGTGGATGGGACGGATTCGGACATAGACAGT with the nucleotide at either position 4, 14, 20, or 24 replaced by C6dT. The first strand additionally contained a biotinylated poly-A tail for surface immobilization for room-temperature experiments that were not discussed here. Modified oligonucleotides were purchased from Microsynth

AG (Balgach, Switzerland), purified with ion-exchange chromatography and labeled with Alexa Fluor 532 succinimidyl ester (Invitrogen). The two strands were then hybridized to obtain a double-stranded DNA. Since double-helical DNA has a persistence length of about 50 nm, we expect our short DNA constructs (less than 15 nm long) to behave like rigid rods [217].

For the conjugation of streptavidin (Sigma) with biotin-Atto647N (Sigma) both proteins were incubated in Tris-EDTA buffer (pH 7.4) at concentrations of 0.8 M for biotin-Atto647N and 16 mM for streptavidin for two hours at room temperature. After reaction, the solution of bound and unbound biotin was purified using a centrifugal concentrator (Vivaspin 20, Satorius) with a cutoff at 30 kDa. The degree of labeling was determined by UV-VIS absorption spectroscopy using a Nanodrop-2000 (Thermo Scientific). We used Tris-EDTA buffer for blanking and took an extinction coefficient of $\epsilon_{280} = 167,000\,\mathrm{M}^{-1}\mathrm{cm}^{-1}$ at 280 nm. The extinction coefficient for Atto647N is given in the data sheet as $\epsilon_{\lambda} = 150,000\,\mathrm{M}^{-1}\mathrm{cm}^{-1}$ and the correction coefficient as $C_{280} = 0.05$ (AttoTec GmbH). The degree of labeling (DOL) was determined to be about DOL = 4.

CitA PASc (residues 200 – 309) from *Geobacillus thermodenitrificans* was modified by site-directed mutagenesis to carry a C-terminal cysteine residue (N308C) which could be used for dye conjugation. Uniformly ^{15}N-labeled CitA PASc N308C was expressed in *Escherichia coli* BL21(DE3) cells in M9 minimal medium with ^{15}N-ammonium chloride as nitrogen-source. After induction with IPTG, cells were incubated in a shaking culture for 5 hours at 30°C before harvesting. Cell pellets were re-suspended in lysis buffer (20 mM Tris · HCl pH 7.9, 300 mM NaCl, 10 mM imidazole, 0.5 mM phenylmethylsulfonylfluoride (PMSF)) and ruptured by sonication. PASc N308C protein was collected via immobilized metal affinity chromatography on Ni-NTA resin (Qiagen). The N-terminal His-tag was cleaved by incubating the protein with TEV protease. Cleaved PASc N308C was reloaded onto Ni-NTA resin and the flow-through of the resin dialyzed against size exclusion chromatography (SEC) buffer (20 mM Na-phosphate pH 6.5, 150 mM NaCl). The final SEC purification was performed with a Superdex 75 16/60 column (GE Life Sciences). The purified PASc N308C sample was dialyzed against phosphate buffered saline (PBS), pH 7.4 and subsequently concentrated to 200 μM. For the conjugation of the PASc domain with Atto647N-maleimide (AttoTec GmbH) we followed the standard protocol (AttoTec GmbH) and removed unbound dye by gel filtration on a SD75 10/30 column. The protein was then exchanged to 10 mM Tris buffer at pH 7.4 and subsequently spin-concentrated. We estimated the concentration by UV-VIS spectroscopy to be about 15 μM using an extinction coefficient of $\epsilon_{280} = 38,400\,\mathrm{M}^{-1}\mathrm{cm}^{-1}$ at 280 nm. The DOL for the PASc domain was determined to be about DOL = 1.3.

A solution for spin-coating was prepared by mixing 90 μl of Tris-EDTA buffer, 20 μl polyvinyl alcohol (PVA) (10 %w, steril filtered, degassed), 10 μl Trolox (20 mM) in DMSO and 10 μl Streptavidin-Biotin Atto647N (aliquots from −20 °C resuspended in 1 ml Tris-EDTA buffer, then diluted 1:100 in Tris-EDTA) or PASc stock solution (aliquots from

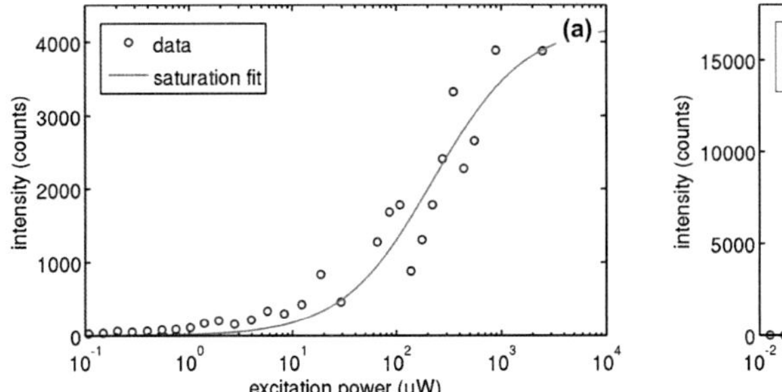

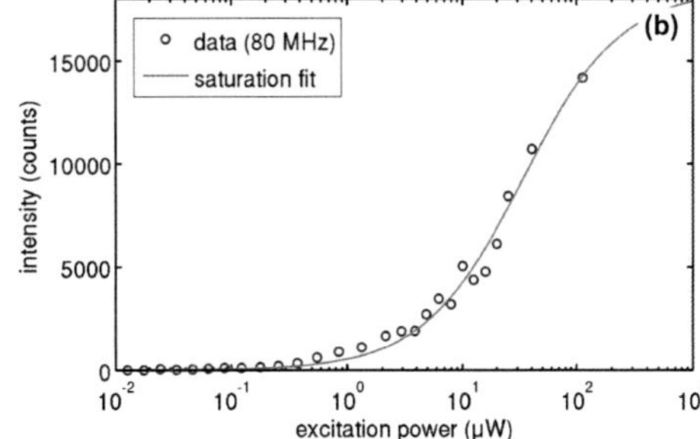

Figure 3.1: Saturation measurements with cw and pulsed excitation at cryogenic temperature. Saturation curves of single Alexa Fluor 488 molecules with cw **(a)** and pulsed **(b)** excitation recorded at 4 K.

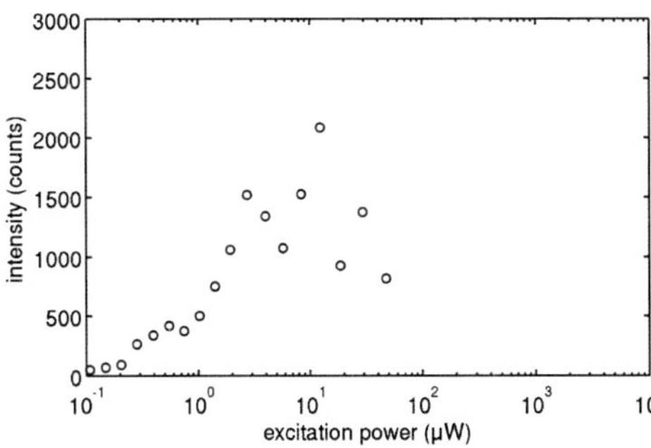

Figure 3.2: Room-temperature saturation measurements with cw excitation. Example saturation curve of single Alexa Fluor 488 molecules with cw excitation recorded at room temperature.

−80 °C, then diluted 1:10,000 in Tris-EDTA), respectively. 5 μl of this solution was spin-coated on the cover slip (10 s at 1000 rpm, 60 s at 3000 rpm), and the sample was immediately transferred to the cryostat and cooled down.

3.3 Results and discussion

3.3.1 Saturation measurements

Typical saturation measurements for single Alexa Fluor 488 molecules at cryogenic temperature are shown in Fig. 3.1 for cw and pulsed excitation. The molecules are excited at a wavelength of 488 nm and the fluorescence is recorded in confocal configuration. The fluorescence rate is fitted to the equation

$$R(I) = R_\infty \frac{I/I_S}{1 + (I/I_S)} \quad , \tag{3.1}$$

where I_S is the saturation intensity and R_∞ the fully saturated emission rate [102]. At

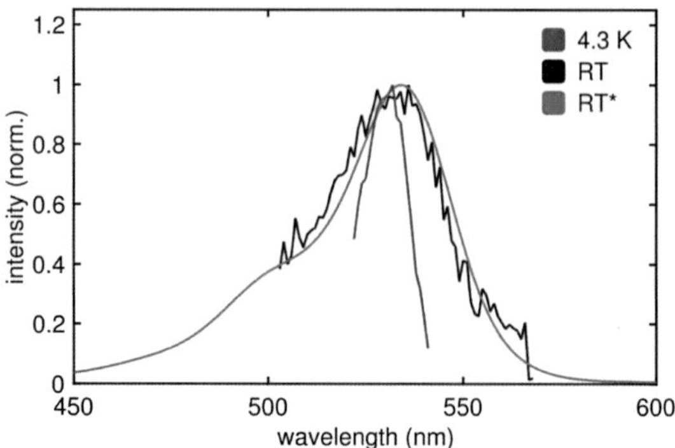

Figure 3.3: Ensemble fluorescence excitation spectra of Alexa Fluor 532. Spectrum of Alexa Fluor 532 recorded using the supercontinuum laser as excitation source at $T = 4$ K (red), room temperature (blue) and comparison to the literature absorption spectrum in solution (gray).

room temperature, the molecules show photobleaching before the intensity ramp is completed, which makes the fitting procedure biased towards very low I_S and very high R_∞ (see Fig. 3.2).

At cryogenic temperatures, statistical values for the saturation intensity and the saturated emission rate can be obtained. For cw excitation, the fitting procedure yields $R_\infty = 7.3 \pm 1.0 \cdot 10^3$ cps and $I_S = 150 \pm 50\,\mu$W, and the corresponding values at pulsed excitation are $R_\infty = 17.5 \pm 1.3 \cdot 10^3$ cps and $I_S = 110 \pm 30\,\mu$W. For subsequent confocal measurements, we selected an excitation power at about $I_S/2$ to ensure an optimal balance between brightness of the fluorescent dye and low background due to non-saturating effects such as luminescence from the substrate.

3.3.2 Ensemble and single molecule spectroscopy

Figure 3.3 shows ensemble absorption spectra of Alexa Fluor 532 at cryogenic temperature (red) and at room temperature (blue). The spectral width of the absorption spectra at low temperature is about FWHM = 15 nm, a factor of two to three narrower than the room temperature spectrum. The spectra were obtained by scanning the wavelength of the supercontinuum laser as excitation source while recording the fluorescence from a wide-field illuminated area of about 30 μm diameter with the SPAD. The excitation light was removed with a 570 nm long pass filter in the detection path. We also plot a literature absorption spectrum of Alexa Fluor 532 in solution for comparison (gray).

We also conducted single molecule emission spectroscopy of Alexa Fluor 532 at cryogenic temperature using a grating spectrometer (see Fig. 3.4). Panels (a) and (c) display two examples of single molecule spectra from the same sample. All molecules we observed have narrower linewidths compared to room temperature. But the linewidths and line shapes of individual molecules at cryogenic temperatures are very diverse even in the same sample. The molecule shown in panel (a) is narrower than at room temperature by a factor of two or three, the molecule shown in panel (c) clearly shows a zero phonon line and the phonon sideband for the 0-0 transition at about 550 nm as well as the 0-1 transition at 590 nm. The branching ratio or Debye-Waller factor was determined to be 28 % in the case of panel (c). Even though the linewidth of the zero phonon line of this molecule is below the resolution of our grating spectrometer, it is most likely far from being lifetime-limited. The strong variation in the spectral linewidths indicates the heterogeneity in the coupling between the molecule and the host matrix.

Fig. 3.4b and d show the temporal evolution of the emission spectra of the corresponding molecules. In panel (d), one notes recurrent shifts of the zero phonon line. Panel (e) shows a time trace of the central position of the zero phonon line of the molecule from (d). The molecule shows both fast and slow spectral diffusion covering a spectral range of about one nanometer. There seems to be a meta-stable spectral position at about 548.25 nm to which the molecule goes back to several times.

The narrowing of the spectral linewidths at cryogenic temperature can be used to perform co-localization by spectral selection. By scanning a narrow-band laser across the inhomogeneous broadening, the molecules can be excited sequentially one after the other. In our case, the inhomogeneous broadening is on the order of $10 - 20$ nm corresponding to about $10 - 20$ THz at a center wavelength of 550 nm. At a homogeneous linewidth of 0.2 THz, about 100 molecules could be identified and subsequently resolved. Compared to lifetime-limited linewidths of about 20 MHz, these are four orders of magnitude fewer molecules but it still constitutes a significant improvement over room temperature. It is worth mentioning that this increased spectral coding ability at cryogenic temperature can also be used to extend the number of color channels in multicolor imaging. At room temperature it becomes difficult to perform labeling with more than five colors. The reason for this is mainly the ratio between the spectral width of the emission lines at room temperature and the available spectral window in the visible regime. It also means that fluorophores with desirable properties must be synthesized for each desired wavelength of emission. At cryogenic temperature, on the other hand, one can utilize the inhomogeneous broadening of one species of fluorophore for a multitude of color channels.

3.3.3 Effect of temperature on photobleaching

We also investigated the influence of the temperature on the photophysics of organic dye molecules. In order to do so, we performed measurements on streptavidin proteins labeled with four Atto647N dyes at room temperature, liquid nitrogen temperature (77 K), 40 K and liquid helium temperature.

Figure 3.5 shows the average number of detected photons per fluorophore in such a conjugate for the four different temperatures. The photon yield is clearly the highest at liquid helium temperature with more than 220,000 average detected photons. At higher temperatures this photon yield is lower, in the case of 40 K and 77 K about $2 - 3$ times lower than at liquid helium temperature.

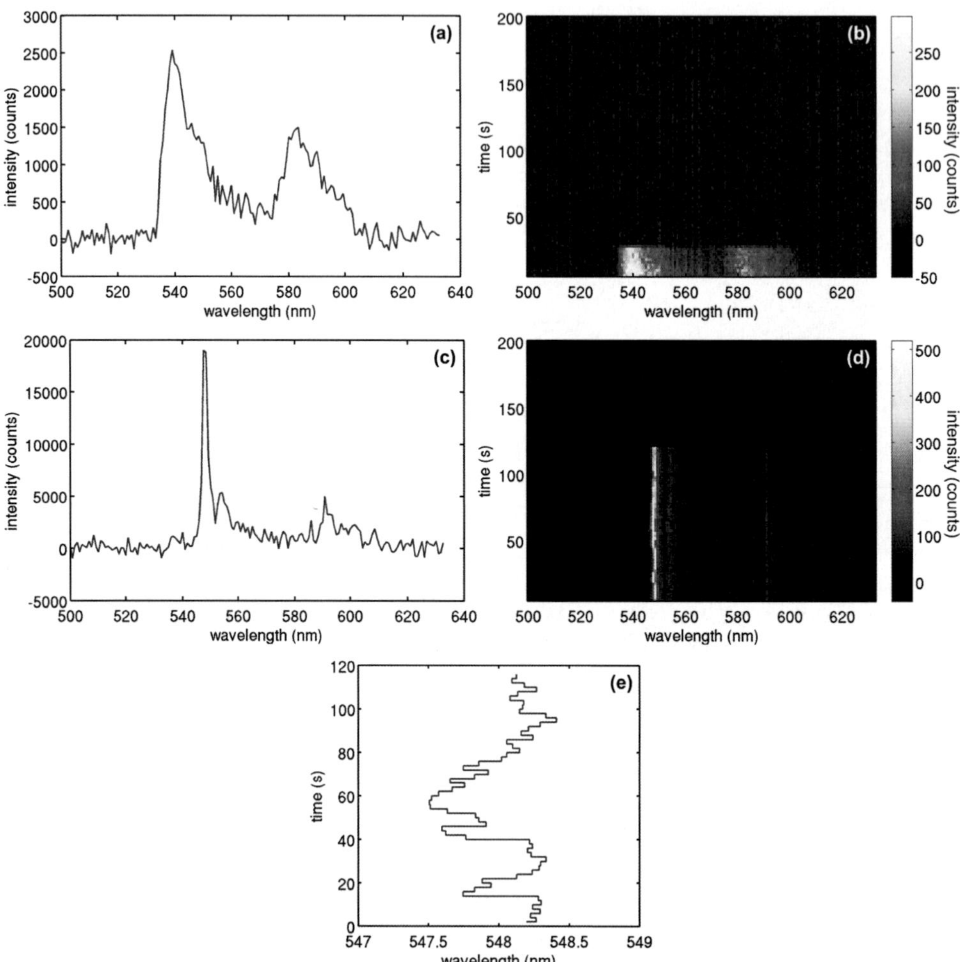

Figure 3.4: Fluorescence emission spectra of single molecules. Emission spectrum of two single Alexa Fluor 532 at low temperature **(a,c)** and time evolution **(b,d)**. **(e)** The time trace of the zero phonon line position from (d) shows spectral diffusion.

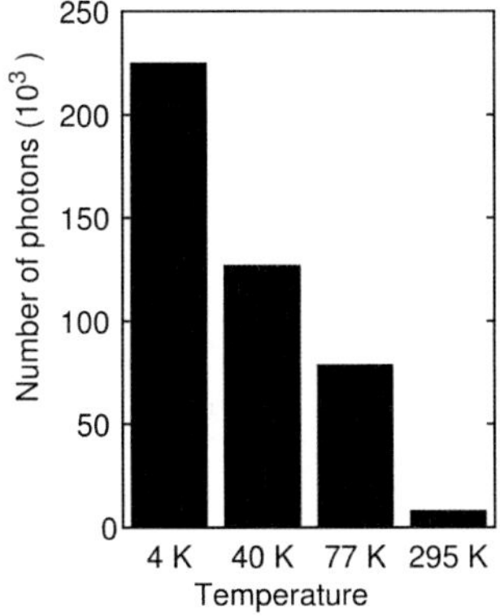

Figure 3.5: Detected photon yield per fluorophore at different temperatures. Number of detected photons per fluorophore in the streptavidin-Atto647N conjugate for three different cryogenic temperatures and room temperature.

3.3.4 Photoblinking of single molecules

In order to study the photoblinking of single molecules, we extracted the on-times and off-times of individual Alexa Fluor 532 molecules from wide-field image stacks, recorded at a frame rate of 20 Hz. The signals from the different frames are connected under the assumption that signals in different frames belong to the same molecule when the position is within a given distance. The connection algorithm also allows for a dead time between frames thereby taking photoblinking into account. In the case of frames where no signal could be identified, the value is set to zero.

To investigate this blinking behavior more quantitatively, we computed single molecule histograms for the off-times and for the on-times from fluorescence traces. Figure 3.6a shows an example of such a single molecule histogram for an Alexa Fluor 532 molecule. The straight line in the double-log plot over about three decades indicates that the on-times and the off-times follow a power law distribution,

$$P(\tau_{\mathrm{on/off}}) = P_0 \times \tau_{\mathrm{on/off}}^{-m_{\mathrm{on/off}}}, \tag{3.2}$$

with $m_{\mathrm{on/off}}$ being the power law exponent. We determine the exponents $m_{\mathrm{on/off}}$ by maximum likelihood estimation. Figure 3.6b shows a histogram for the determined power law coefficients for the on-times and off-times. Due to parallel recording of many fluorescence traces with a wide-field microscope, we can afford to discard fluorescence traces where the fitting algorithm does not yield a result for the power law exponent with confidence bounds of at least 95 %. In the case of Alexa Fluor 532 we find a distribution of m_{on} with a clear peak at about $m_{\mathrm{on}} = 1.7$ and we can observe that the distribution has a very broad width. This value is similar to values reported from ensemble histograms of comparable dye molecules (ATTO 565 and rhodamine 6G) at room temperature [111,113].

Furthermore, we see similar behavior for the off-times power-law exponents: The

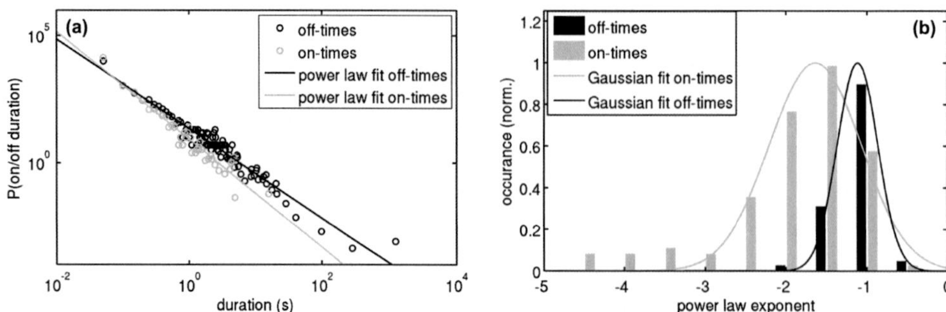

Figure 3.6: Photoblinking of single Alexa Fluor 532 at cryogenic temperature. (a) On/off-times histogram of a single Alexa Fluor 532 molecule and fit result to a power law distribution. **(b)** Histogram of the power law exponents for the on-times and off-times.

distribution has a peak at approximately $m_{\text{off}} = 1$. This value is considerably lower compared to values from the afore mentioned studies ($m_{\text{off}} = 1.8 - 2.1$) [111, 113]. A lower power-law exponent means that statistically more longer off-periods occur. This is consistent with a charge tunneling model developed to describe fluorescence intermittency in semiconductor nanocrystals [218]. Here, a charge is transferred from the molecule to a trapped state in the polymer matrix. The following charge recombination can only happen via charge tunneling from the charge trap to the molecule cation. Since the tunneling rate is exponentially dependent on the distance, the off-time durations follow power law statistics [110]. In this picture, the exponent m_{off} is defined by the tunneling barrier which the charge has to overcome. The lower power-law exponent therefore suggests that the tunneling barrier is higher for the investigated molecules at cryogenic temperature compared to room temperature. As a consequence, the dye molecules stay on average longer in this long-lived dark state at 4 K.

We also recorded time-tagged intensity traces of single Alexa Fluor 532 molecules using confocal detection. Figure 3.7a and b show examples of time traces obtained by binning the arrival times of the photons to 10 ms bins at different zoom levels. From the time-tagged photon stream we also calculated the second-order autocorrelation $g^{(2)}(\tau)$ as function of the time lag τ according to the equation

$$g^{(2)}(\tau) = \frac{\langle I(t)I(t+\tau)\rangle}{\langle I(t)\rangle^2} \quad , \tag{3.3}$$

where angle brackets denote time averaging over a long time $T \gg \tau$ and $I(t)$ is the intensity within a small bin with central time t. It has been already pointed out that correlation functions offer much more reliable analysis of photoblinking data [219]. First, they do not rely on the definition of "on" and "off" by means of a threshold which choice may affect the conclusions. Second, an insufficient number of orders of magnitude or

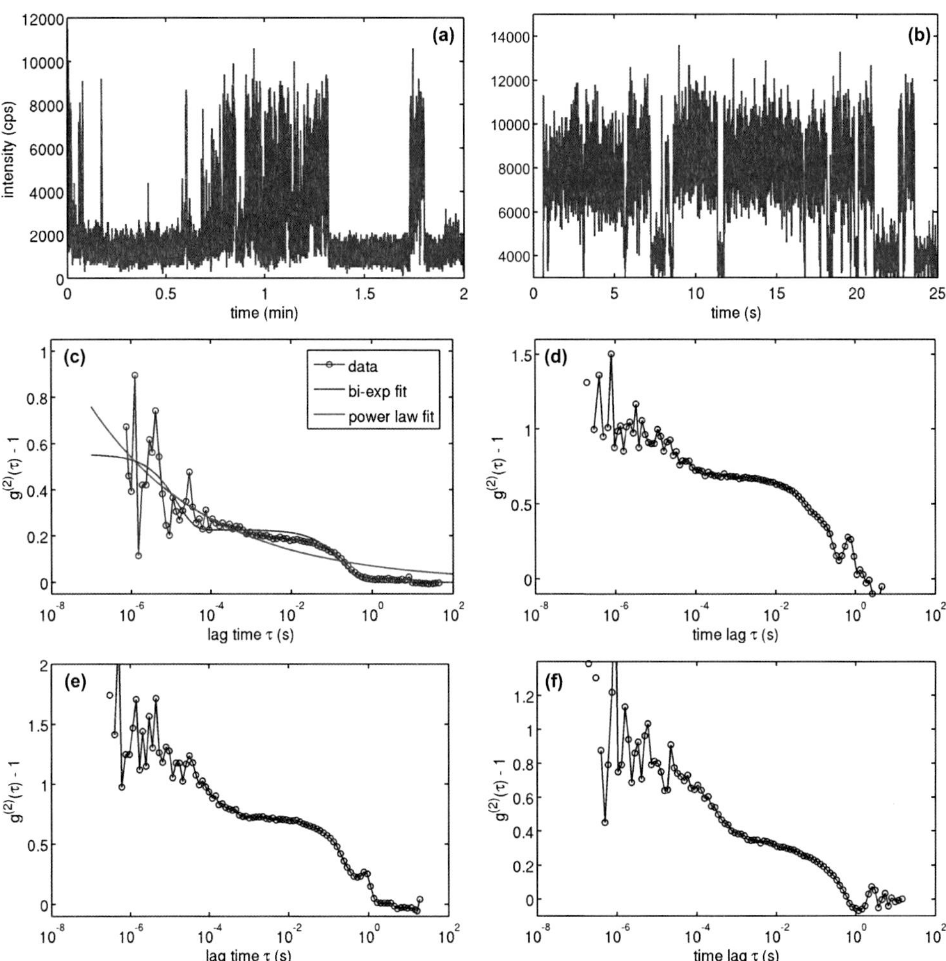

Figure 3.7: Autocorrelation measurements computed from TCSPC data. Examples for time traces **(a,b)** from TCSPC data (10 ms bins) and second order autocorrelation functions **(c–f)** ($g^{(2)}$ function) of single Alexa Fluor 532 molecules at 4 K.

too much noise in an on/off-time histogram may cause confusion of a multiexponential decay with a power-law distribution.

Figure 3.7c–f shows autocorrelation functions of the emission from single Alexa Fluor 532 molecules measured at cryogenic temperatures. Interestingly, the plots show a clear deviation from a power-law behavior and a single-exponential distribution (Fig. 3.7c, red line), and the data are better described by a bi-exponential function. The $g^{(2)}(\tau)$ function of the molecule shows two characteristic time scales. One shorter photoblinking happens on the order 10 µs and a second fluorescence intermittency takes place at much longer times on the order of 100 ms. Such autocorrelation functions can be explained by a four-state model [219] which comprises of two two-level blinking entities of unequal brightness at different intermittency rates, between which the systems switches at a third rate. We attribute the short-time fluorescence intermittency to triplet state dynamics. In our experiments the characteristic time is longer than what similar dye molecules show at ambient conditions. A similar effect has been observed when molecular oxygen has been removed from the solution [108]. The long time scale photoblinking may be caused by a charge trapping mechanism or by the molecule undergoing chemical changes [110].

3.3.5 Photoblinking of multi-fluorophore conjugates

We also investigated the photoblinking behavior of single Atto647N molecules at cryogenic temperatures. An exemplary intensity trace of a single Atto647N molecule and the ensemble photoblinking statistics is shown in Fig. 3.8. To test the influence of Atto647N binding to a protein, we compared the blinking times of free Atto647N, Atto647N-biotin ligands and constructs where a protein was labeled with two Atto647N (PAScWT) and with four Atto647N (streptavidin) fluorophores (see Fig. 3.8c).

While the on-times of free Atto647N-NHS are very similar to those when the fluorophore is attached to the small biotin protein with a molecular weight of $M_w = 244$ Da, we found significant differences when the fluorophores were fused to larger proteins. Atto647N fluorophores when coupled to streptavidin ($M_w = 52.8$ kDa) or the cytosolic PAS domain from the histidine kinase CitA of *Geobacillus thermodenitrificans* (PAScWT, $M_w = 12.2$ kDa) showed shorter on-times by a factor of about five compared to free Atto647N fluorophores dispersed in the polymer matrix (see Fig. 3.8d).

Moreover, we compared the fluorescence intensity for single Atto647N fluorophores with individual Atto647N being part of the four fluorophores attached to streptavidin. In this analysis the Atto647N coupled to streptavidin showed on average a lower count rate indicating a quenching effect (see Fig. 3.8e).

3.3.6 Quenching of coupled fluorophores

Using the DNA ruler samples with two fluorophores at 10 nm, 7 nm, 5 nm, and 3 nm separation, we also investigated the fluorescence quenching for different fluorophore

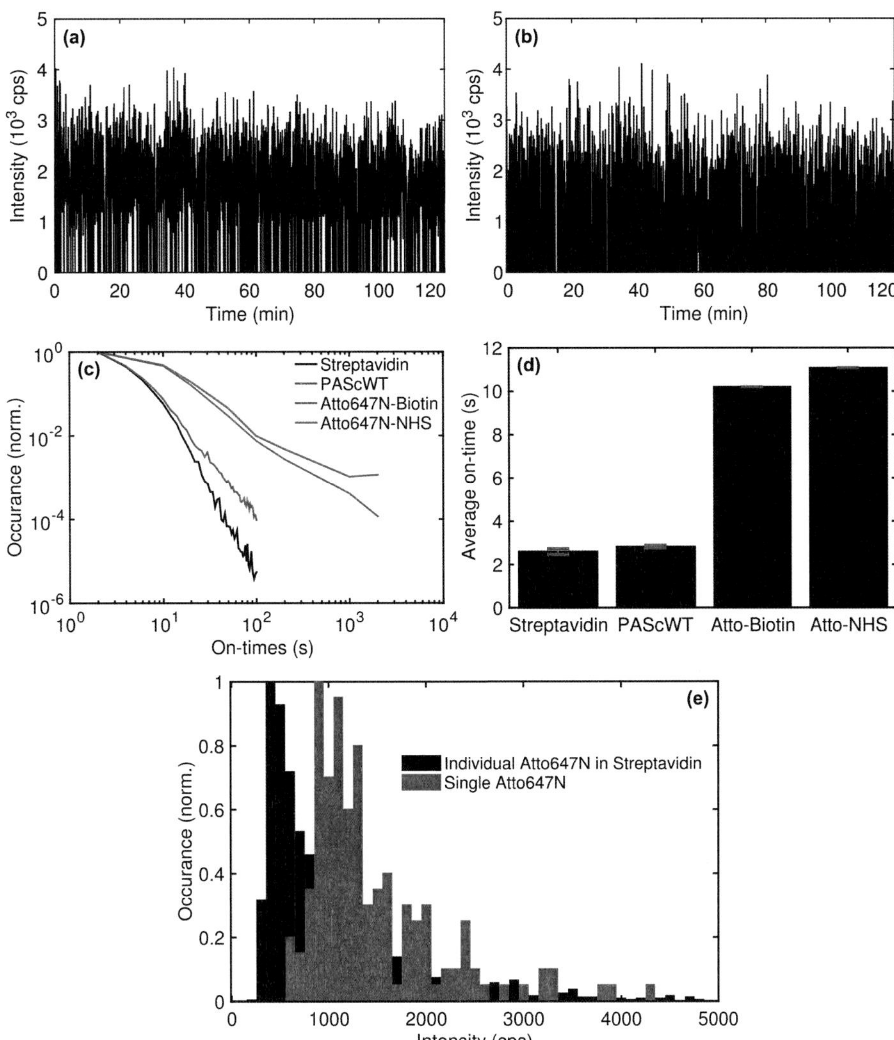

Figure 3.8: Single molecule photophysics of Atto647N at liquid helium temperature. Intensity time trace for Atto647N-NHS **(a)** and Atto647N-biotin **(b)**. **(c)** On-time photoblinking statistics for single Atto647N-NHS (green) and single Atto647N-Biotin (blue), for individual Atto647N coupled via biotin to streptavidin (black) and individual Atto647N coupled to PASc (red). **(d)** Histogram plot of the characteristic on-times for the individual fluorophores in the streptavidin and PASc system as well as single Atto647N fluorophores coupled to biotin and as bare fluorophore. **(e)** Count rate for single Atto647N fluorophores coupled to streptavidin (black) and for the bare fluorophores (red).

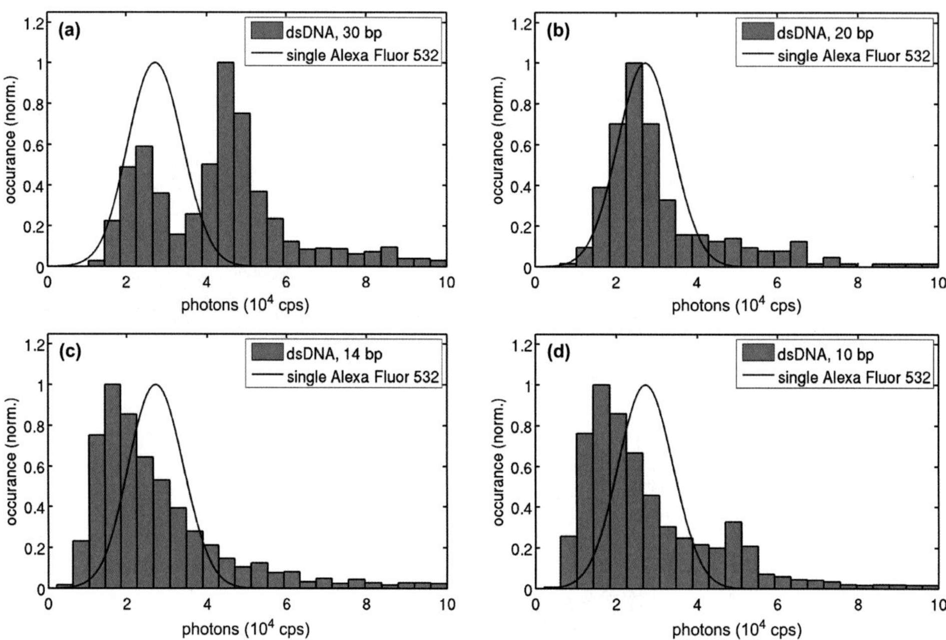

Figure 3.9: Intensity histograms from wide-field recordings of double-stranded DNA molecules labeled with two Alexa Fluor 532 molecules (room temperature). (a) 30 bp, **(b)** 20 bp, **(c)** 14 bp, **(d)** 10 bp fluorophore separation.

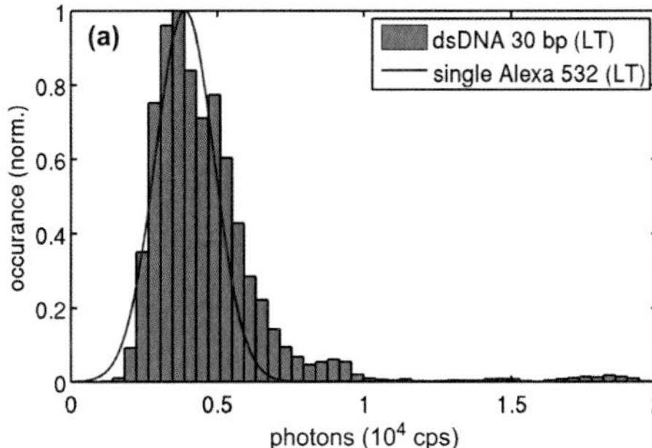

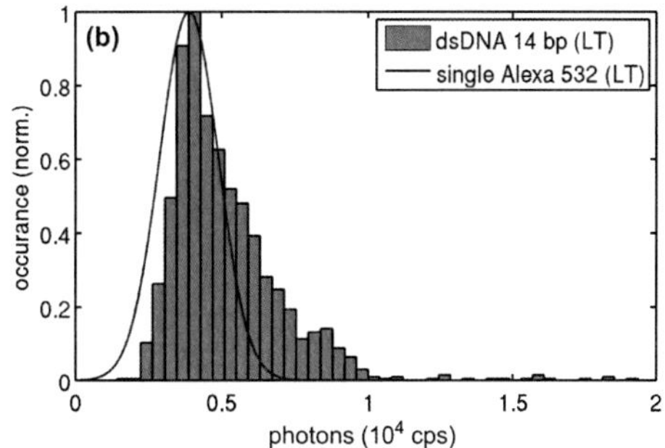

Figure 3.10: Intensity histograms of wide-field recording of double-stranded DNA molecules labeled with two Alexa Fluor 532 molecules (low temperature). (a) 30 bp and **(b)** 14 bp fluorophore separation.

separations. Figure 3.9 shows histograms of the intensities of identified spots from a series of wide-field image stacks recorded at room temperature and for decreasing separation from panel (a) to (d). For comparison, the histogram of single Alexa Fluor 532 molecules is also plotted into each of the histograms. At 10 nm separation, the single molecule peak overlaps with the first peak in the histogram and the second peak, corresponding to two independently emitting fluorophores, lies at double the intensity. The integral peak ratio is here determined by photoblinking and photobleaching. Already at 7 nm separation, the higher intensity peak is beginning to vanish and the first peak is starting to shift to lower intensities indicating quenching due to the decreasing separation and therefore increasing coupling between the two fluorophores.

Interestingly, when we do the same kind of analysis on data obtained at cryogenic temperatures (Fig. 3.10), we do not observe this quenching behavior. At cryogenic temperatures, the fluorophores exhibit a different photoblinking behavior with a much lower yield of molecules showing two-step photoblinking or emitting at the double intensity. Furthermore, the fraction of molecules showing fluorescence at the double intensity seems to be independent of the fluorophore separation.

3.3.7 Fluorescence lifetime of multi-fluorophore systems

In order to further investigate this coupling phenomenon, we recorded time-tagged intensity traces from DNA rulers with different separation as well as proteins that were labeled with multiple fluorophores at very close distance. From these time-tagged traces, we can calculate the time evolution of the fluorescence lifetime by computing lifetime histograms from 500 or 1,000 photons per histogram. This number of photons is sufficient to determine the fluorescence lifetime with high confidence [220]. Lifetime values where the uncertainty of the fit is larger than 0.25 ns are discarded.

Figure 3.11a shows the intensity time trace and fluorescence lifetime trace of an

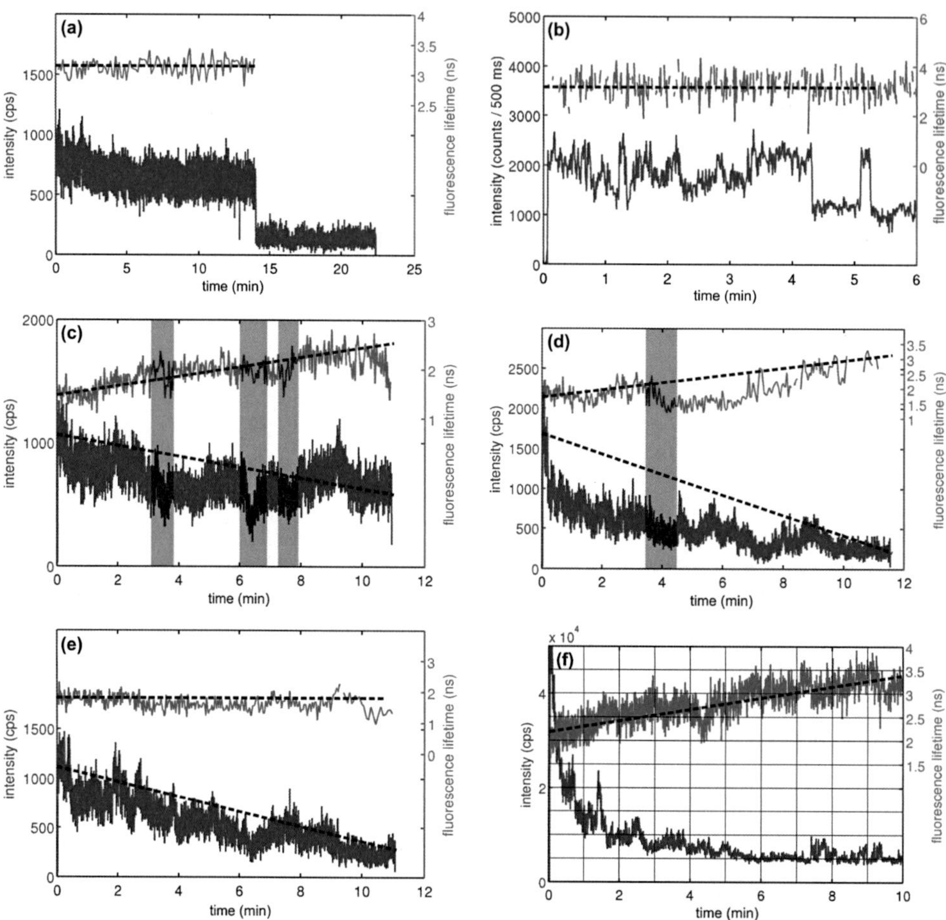

Figure 3.11: Fluorescence lifetime traces recorded at cryogenic temperature. Shown is the intensity time trace (blue) and the time evolution of the fluorescence lifetime (green). Dashed lines serve as guide to the eye. **(a)** Single Alexa Fluor 532 molecule (4 K). **(b)** DNA ruler with two Alexa Fluor 532 molecules at 30 bp separation and **(c–e)** three examples for 10 bp separation. Shaded areas mark regions of correlation between intensity time trace and fluorescence lifetime trace. **(f)** Laminin protein labeled with DOL = 14 (average labeling density) Alexa Fluor 532 molecules.

individual Alexa Fluor 532 molecule. In panels (b) and (c) of Fig. 3.11 we show the same type of data also for DNA rulers with 10 nm and 3 nm separation, respectively. While for the longer separation the fluorescence lifetime stays constant and, in particular, does not show any kind of correlation with the signal intensity of the two photoblinking fluorophores, there is a clear correlation for the shorter distance shown in panels (c–e). At higher intensities, both fluorophores are in a bright state and the fluorescence lifetime is reduced. We marked some regions in Fig. 3.11c and d that exhibit clear correlations of the intensity time trace and the fluorescence lifetime trace. Interestingly, this effect is not always present to the same extent as can be seen in the third example shown in Fig. 3.11e. This variation in the amount of coupling for the same species of DNA ruler can be explained by different mutual orientations of the fluorophores.

Fig. 3.11c,d. The correlation between fluorescence intensity and fluorescence lifetime is more dramatically demonstrated in panel (f), where we show the time trace of a single laminin protein that is labeled with Alexa Fluor 532 at an average labeling density of DOL = 14. As more and more fluorophores go to dark states, the initially strongly reduced fluorescence lifetime gets closer to the single molecule fluorescence lifetime (see also Fig. 3.12). These results are in accordance with room-temperature observations of an increase in the fluorescence intensity from antibodies labeled with an increasing number of fluorophores [221].

Luchoswki and coworkers showed that excessive labeling leads to fluorescence intensity decrease and an average fluorescence lifetime decrease [221]. Their explanation is based on non-fluorescent traps in the process of homo-FRET that can act as a non-radiative sink for the excited state energy. It is worth noting that the fluorescent dye used in that study (Seta-670) has a relatively low fluorescence quantum yield of $\Phi_D = 0.07$ but a similar fluorescence lifetime of $\tau_D = 2.4$ ns compared to the dye molecules used in this work. The fluorescence quantum yield of the donor can play an important role for the degree of coupling between two fluorophores. The expression for the excitation energy transfer rate reads (omitting prefactors) [222]

$$k_{DA} \propto \frac{1}{n^4} \frac{\Phi_D}{\tau_D} \frac{\kappa^2}{|r_{DA}|^6} \int_0^{\infty} \frac{d\omega}{\omega^4} I_D(\omega) \sigma_A(\omega) \quad , \tag{3.4}$$

where n is the refractive index of the medium, Φ_D denotes the fluorescence quantum yield of the donor in absence of the acceptor, and τ_D the fluorescence lifetime of the excited state of the donor in absence of the acceptor. A lower quantum yield translates into a lower efficiency of excitation energy transfer. On the other hand, possible excitation energy transfer may be hampered by the narrower spectral line widths at cryogenic temperatures which reduce the spectral overlap integral. It is worth mentioning that the fluorescence lifetime of fluorophores and their quantum yield are not independent of each other but inherently connected. The fluorescence quantum yield is defined as the ratio of the radiative excited state decay rate and the sum of all rates of excited state

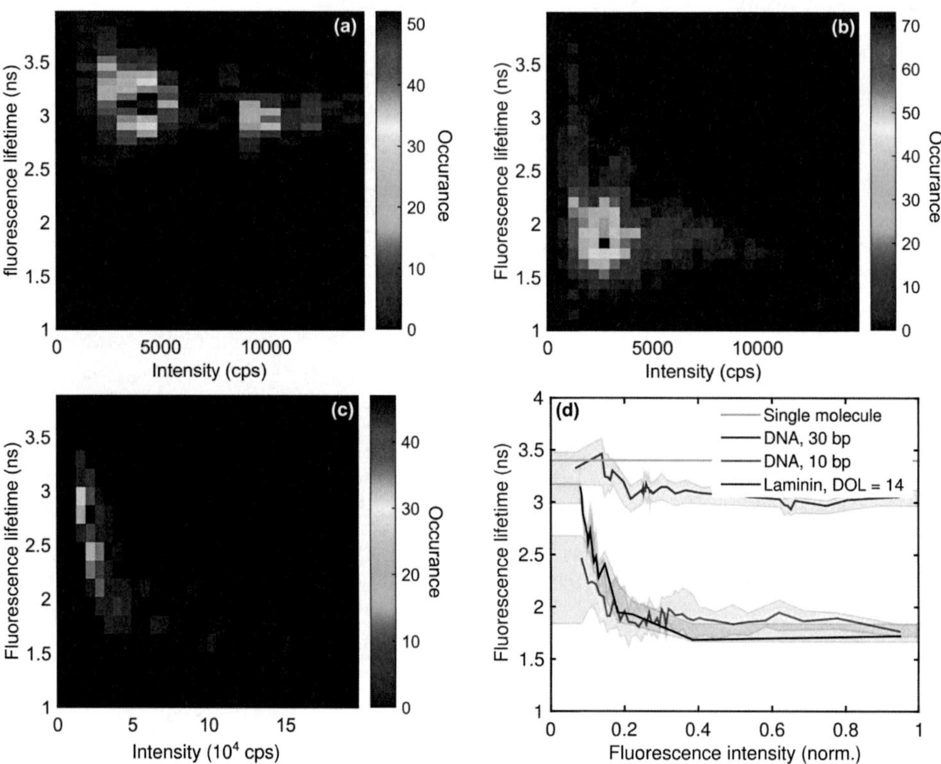

Figure 3.12: Statistics of the cryogenic fluorescence lifetime traces. Fluorescence lifetime / fluorescence intensity correlations for a DNA ruler with 10 nm separation **(a)**, 3 nm separation **(b)**, and labeled laminin **(c)**. **(d)** Fluroescence lifetime as a function of the fluorescence intensity for single molecules, the two DNA constructs and the labeled laminin. This panel contains the same data as panels (a), (b) and (c).

decay of the molecule [211].

3.3.8 Photoblinking of fluorophores with different quantum yields

In order to investigate the influence of the quantum yield on the photophysics of protein-fluorophore constructs, we recorded wide-field image stacks of streptavidin proteins labeled with three different fluorophores. We computed histograms for the intensities and identified levels of intensities for conjugates of streptavidin with Alexa Fluor 488 with a quantum yield at room temperature of $\Phi_F = 0.92$, Alexa Fluor 532 with $\Phi_F = 0.61$ and Alexa Fluor 647 which has $\Phi_F = 0.33$. It is important to note that the investigated molecules also differ in fluorescence lifetime. The ratio of fluorescence quantum yield to fluorescence lifetime which directly influences the degree of excitation energy transfer changes only moderately and is the highest for Alexa Fluor 488 $\Phi_F/\tau_F \approx 0.2$ and the lowest for Alexa Fluor 647 at $\Phi_F/\tau_F \approx 0.15$. Note that for this calculation we used experimental values for the fluorescence lifetimes recorded at cryogenic temperatures. These values are different from the literature values which were obtained at room temperature and in solution. At the same time, we use literature values for the fluorescence quantum yields since determining the quantum yield of single fluorophores in cryogenic measurements is not a straightforward exercise. This introduces an error in our calculation that we cannot estimate. Recently, a planar metallo-dielectric antenna that allows to collect more than 99 % of the emitted photons from single emitters has been experimentally demonstrated [223]. Using such an antenna would allow it to determine the quantum yield of single molecules with high precision by driving the molecules in saturation and monitoring the emitted fluorescence rate.

Figure 3.13a–c displays intensity histograms of the three types of dye molecules. In these histograms we have also taken into account the occurance of frames at each level. There is a clear trend that it becomes more likely to observe higher intensity levels with decreasing quantum yield. In the case of Alexa Fluor 488 (Fig. 3.13a), the dye molecule with the highest quantum yield $\Phi_F = 0.92$, only two of the peaks are pronounced, whereas in the case of Alexa Fluor 647 ($\Phi_F = 0.33$) all four intensity peaks are clearly visible (see Fig. 3.13c).

Figure 3.13d displays a histogram for the identified levels from which photobleaching in the fluorescence traces occured. For all three investigated dye molecules, photobleaching is most likely to occur from the lowest level of emission. Figure 3.13e presents a histogram for the number of identified levels per extracted intensity trace. A similar kind of phenomenon can be observed as in Fig. 3.13a–c. It becomes more likely to observe higher intensity levels with decreasing quantum yield. In the case of Alexa Fluor 488 ($\Phi_F = 0.92$) the most probable identified level of intensity is 2. The fluorophore with the lowest quantum yield, Alexa Fluor 647 with $\Phi_F = 0.33$, on the other hand, has a peak at 4 identified levels in the histogram.

Table 3.1 summarizes both the properties of the investigated dye molecules and our

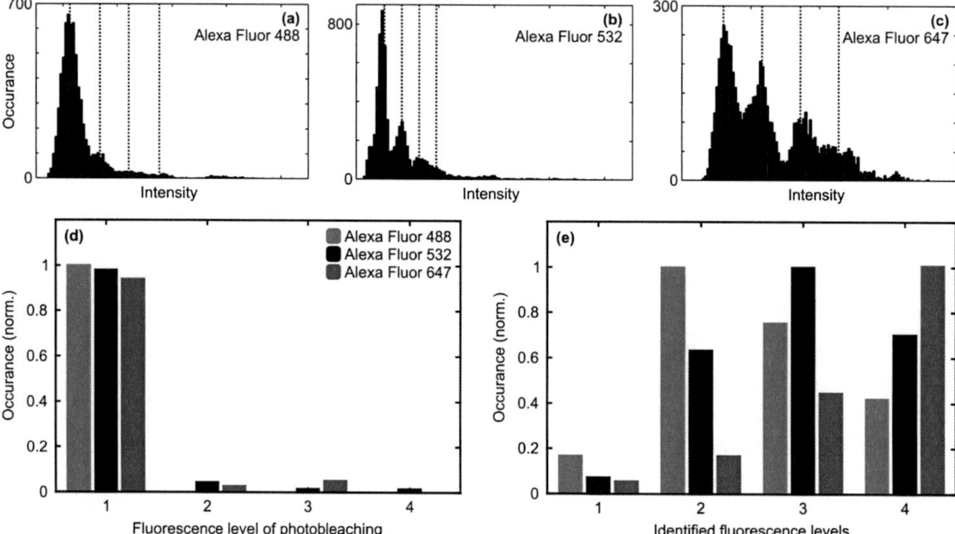

Figure 3.13: Streptavidin conjugated with fluorophores of different quantum yield. (a–c) Histograms for the intensities of all recorded frames for the three investigated fluorophores Alexa Fluor 488, Alexa Fluor 532 and Alexa Fluor 647. **(d)** Histogram for the identified fluorescence levels from which photobleaching occured. **(e)** Histogram for the number of identified fluorescence levels.

Fluorophore	family	Φ_F	τ_F	τ_F (LT)	I (LT)	T (LT)
Alexa Fluor 488	rhodamine	0.92 *	4.1 ns *	4.5 ns	+	+
Atto 488	rhodamine	0.80 *	4.1 ns *		+	+
Alexa Fluor 532	rhodamine	0.61 *	2.5 ns	3.4 ns	0	+
Atto 532	rhodamine	0.90 *	3.8 ns *		0	+
Alexa Fluor 647	cyanine	0.33 *	1.0 ns *	2.2 ns	−	+
Atto 647N	carbopyronine	0.65 *	3.5 ns *		+	+

Table 3.1: Summary of the properties of the investigated dye molecules. Φ_F – room temperature quantum yield; τ_F – room temperature fluorescence lifetime; τ_F (LT) – low temperature fluorescence lifetime; I (LT) – fluorescence intensity compared to room temperature; T (LT) – survival time compared to room temperature. Asterisks denote literature values from the manufacturer datasheets.

qualitative results obtained from analyzing all data from cryogenic measurements. In general, we observed longer fluorescent lifetimes at cryogenic temperatures than the room-temperature literature values (in solution). Some molecules become brighter at cryogenic temperatures while others have approximately the same brightness or in one case it even becomes darker. We always observed that the survival time before photobleaching increases, this increase can be between ten to more than one hundred times.

3.4 Summary

We studied the photophysical properties of a selection of organic dye molecules as well as multi-fluorophore systems embedded in a polymer matrix at the single-molecule level and at cryogenic temperatures.

In the experiments on individual single fluorophores, we made the following observations:

- The dye molecules show narrower linewidths at T = 4 K than at room temperature, but there is also a strong variation in the observed linewidths. This indicates that the coupling of the dye molecules to the host matrix is heterogeneous. The fluorophores also exhibit spectral diffusion.

- They have the highest photon yield at liquid helium temperature. At liquid nitrogen temperature (T = 77 K), the photon yield is also increased as compared to room temperature but about a factor of three worse than at T = 4 K. In general, we observed for all fluorophores longer survival times at cryogenic temperatures.

- We observed a power law distribution in the on/off-times histograms and we observed larger power law coefficients for the off-times than has been observed for similar dye molecules at room temperature. This indicates the presence of long-lived dark states which can be on the order of many seconds or even many minutes.

- We computed second-order autocorrelation functions of TCSPC data for single Alexa Fluor 532 molecules and found the signature of a four-state model of unequal brightness, i.e. two characteristic photoblinking times. We found that the triplet-state kinetics happen at slower time scales than typically at room temperature.

We also carried out experiments to study the photophysics of multi-fluorophore systems at low temperature. In this case, we observed the following:

- We investigated the photoblinking behavior of Atto 647N molecules and compared isolated single molecules to the case when multiple molecules are attached to a protein. We observed a shorter average on-time in the case of several molecules close to each other and also a quenching of the fluorescence emission of individual dye molecules when they are part of a protein conjugate.

- We looked at pairs of Alexa Fluor 532 molecules attached to the backbone of a double-stranded DNA. We varied the separation of the two dye molecules between 10 nm and 3 nm and compared cryogenic-temperature to room-temperature measurements. At room temperature, we observed an increased quenching effect with decreasing separation of the dyes. At cryogenic temperatures, however, we did not observe quenching to that extent.

- We observed correlations of changes in the fluorescence lifetime with the photoblinking of fluorescently labeled DNA rulers and proteins. In the case of the DNA rulers, we observed that with decreasing separation of the fluorophores, the fraction of molecules showing an anti-correlation in fluorescence lifetime with fluorescence intensity increases. When performing the same experiment with a protein conjugated to a larger number of fluorophores, this anti-correlation becomes considerably more pronounced.

- We investigated the influence of the fluorescence quantum yield on the photophysics of protein conjugates by performing experiments on streptavidin labeled with three different Alexa Fluor dye molecules with different quantum yields. Here, we observed on average higher levels of intensity in photoblinking when using dye molecules with lower quantum yield. Additionally, these higher levels of intensity also occur more frequently with the lower quantum yield fluorophores.

The photophysics of dye molecules can be complex because they result from the interplay of several effects. In addition to this, different kinds of dye molecules may exhibit different photochemistry. We observed that the concentration of the triplet quencher Trolox influences the photoblinking behavior of Atto 647N molecules. This suggests that even at liquid helium temperature chemistry can have a strong effect. Furthermore, the energy landscape of the host matrix may be intricate. For instance, energy traps can cause long-lived dark states and the refractive index can strongly influence resonance energy transfer. Effects such as charge transfer (Dexter energy transfer [224]) can occur. Also, at too small distances the point-dipole approximation fails [225, 226]. It has also been experimentally observed that through-bond coupling can yield different efficiencies compared to through-space coupling [227]. In this study, terrylene diimide was coupled to perylene diimide using a rigid p-terphenyl spacer to produce a dyade chromophore with defined separation. They observed an increased coupling between the effective donor and acceptor, which was attributed to de-localization of the transition densities over the spacer. Charge transfer along the helices of DNA has also been observed [228].

Fluorophores can have narrow spectral linewidths at cryogenic temperatures and they can be at different spectral positions within the inhomogeneous broadening. In this case, the substantially reduced spectral overlap may hamper efficient coupling, contrary to the common expectation. Furthermore, the transition dipole orientations are frozen, so that the degree of coupling is defined by the mutual orientation of the dye molecules. At the same time, coupling may also be enhanced because of the higher coherence at low temperatures. Moreover, the fluorescence quantum yield clearly has an influence on dipole-dipole coupling but it is not independent of other properties such as fluorescence lifetime. These issues will have to be investigated.

4 Single molecule localization analysis

In this chapter, we describe the single molecule analysis that was used to extract information from the measured data in this work. The electromagnetic field of an emitting point dipole can be computed analytically from Maxwell's equations. The diffraction of this field at a circular aperture can be very well described by the Kirchhoff Vector Approximation. The theory for the point spread function of a dipole emitter at an interface with arbitrary orientation is summarized in Reference [156]. Here, we will mostly follow the notation used in Reference [121].

4.1 Image of a point source

The diffraction of light at the aperture of the microscope objective causes the image of a "point source" such as a fluorescent molecule to appear as a blurred blob on the camera chip (see Fig. 4.1). In other words, the experimental data that we record from a single emitter is a distribution of photon counts n_i in pixel i. The expected photon count E_i in pixel i is

$$E_i(\theta) = Np_i + b^2 \tag{4.1}$$

with $\theta = (N, \mu, s, b^2)$ representing a set of parameters, where N is the total number of photons emitted by the point source across the whole image plane, p_i is the probability

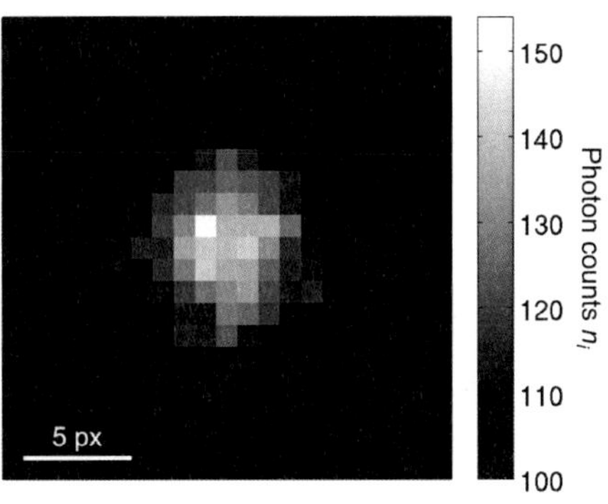

Figure 4.1: Single molecule image. Example of a diffraction-limited spot on the camera chip.

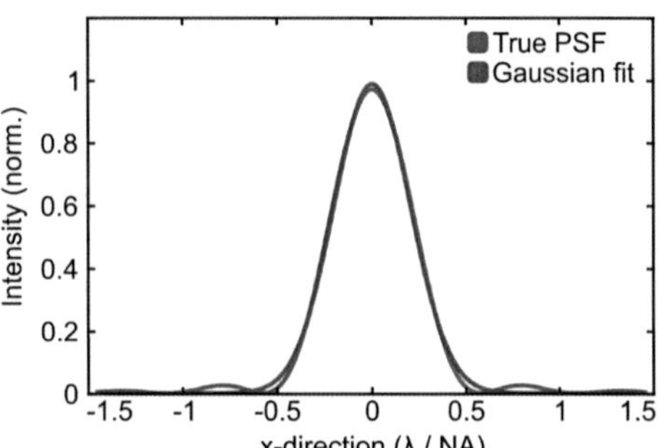

Figure 4.2: Comparison of the true PSF and a Gaussian fit. True PSF of a horizontal dipole at a glass interface for a 100× /0.9 NA microscope objective (see chapter 2 for more details) and a Gaussian fit.

distribution across the image plane (the point-spread function, PSF), and b^2 denotes the background photon count per pixel. The parameter set also contains $\boldsymbol{\mu}$, which represents the coordinates of the point source in x and y direction, and a width parameter of the PSF s. The probability distribution of the PSF is normalized, i.e., $\sum_i p_i = 1$.

If the theoretical description of the probability distribution is correct, then a parameter set θ^* exists such that

$$E_i(\theta^*) = E(n_i) \quad . \tag{4.2}$$

The task of single molecule localization is to determine the true set of parameters θ^* which also contains the expected coordinates $\boldsymbol{\mu}$.

4.2 2D Gaussian fitting

A common approach in single molecule localization is to fit a two-dimensional (2D) Gaussian function to the experimental PSF [119]. A 2D Gaussian, however, is not the true point-spread function. Figure 4.2 plots the true PSF calculated from vectorial theory for the geometry of our second-generation setup (see chapter 2 for details) compared to a fit with a Gaussian function. The true PSF (blue) can be well described by a Gaussian function (red) with a coefficient of determination $R^2 = 0.998$. It slightly deviates from the true PSF shape not exhibiting the shoulders at about $\pm 0.8\,\lambda/\mathrm{NA}$. When allowing for an offset of the Gaussian function and considering it in a window of the width $2\,\lambda/\mathrm{NA}$, the true PSF can be even better approximated [121,229]. It is worth noting that the true PSF also significantly deviates from an Airy distribution. The reason for this are the vectorial diffraction effects that blur the fringes of the Airy distribution [230].

4.2.1 Localization accuracy

The localization accuracy quantifies how close the estimated position lies to the true position. One of the origins of systematic errors causing a limited accuracy is the non-uniformity in the camera pixel response (PRNU) [13]. In our experiments, the camera has a PRNU standard deviation of about 0.5 %. We verified by simulations that this causes a maximal systematic deviation from the true position of less than 0.5 nm. Furthermore, since both emitters are very close and feel the same PRNU landscape on the camera, this systematic error is much further reduced and becomes negligible for our measurements.

Another effect that could affect the localization accuracy is the overlapping PSFs of neighboring entities. This systematic error starts to be insignificant for inter-PSF distances of about nine pixels or more in our configuration. Thus, we avoided samples with too high fluorophore densities.

The strongest systematic localization error is typically due to the emission characteristics of a fixed dipole of arbitrary orientation at an interface. It is generally known that dipole emitters with an inclination angle outside the horizontal plane produce asymmetric PSFs. As a result, a localization method based on a centroid calculation or a 2D Gaussian fit cannot determine the actual position of the dipole emitter [155,156]. Even in the case of rotating dipoles a systematic position error remains when the molecular rotation is partially impaired [231,232]. It has been pointed out that this asymmetry and its associated localization error are much less pronounced if a microscope objective with low numerical aperture is used [156]. To investigate these effects, we calculated artificial PSFs numerically using the Kirchhoff vector approximation following the work of Mortensen and coworkers [121] and references therein.

First, we investigated for the first-generation cryostat configuration if the single-molecule localization is affected by these effects. The calculations were performed for an emission wavelength of 555 nm and a simple generic geometry with one interface between two media of refractive indices $n = 1.0$ and $n = 1.5$ (see Fig. 4.3a). In Fig. 4.3b, we show some examples of calculated PSFs for two different numerical apertures (NA) of the collection optics and different polar angles (θ) of the dipole orientation. It is evident that the degree of asymmetry is much less for NA = 0.75 as compared to the case of high NA. By fitting these generated patterns with 2D Gaussian functions, we compared the true and "measured" positions of the dipole. Figure 4.3c displays the results of the calculated localization error as a function of NA and θ, where the dipole was placed close to the interface (distance $z = 2$ nm) inside the low-index medium, and detection took place through the high-index material assuming an index-matched immersion situation. We find that the systematic localization error is negligible (less than a few Ångströms) if the imaging optics only collects the emission below the critical angle (i.e. NA < 1). However, it can increase to about 13 nm for high NAs and certain dipole orientations. When the detection is through the low index medium (not shown), we find that the systematic localization error is comparable to the case shown in Fig. 4.3c for NA < 1 and is also negligible. In addition, we examined the influence of the dipole-interface distance z, while the dipole was kept in the focal plane at all times (again detecting through the high-index medium). Figure 4.3d shows the result for NA = 0.75, revealing that for

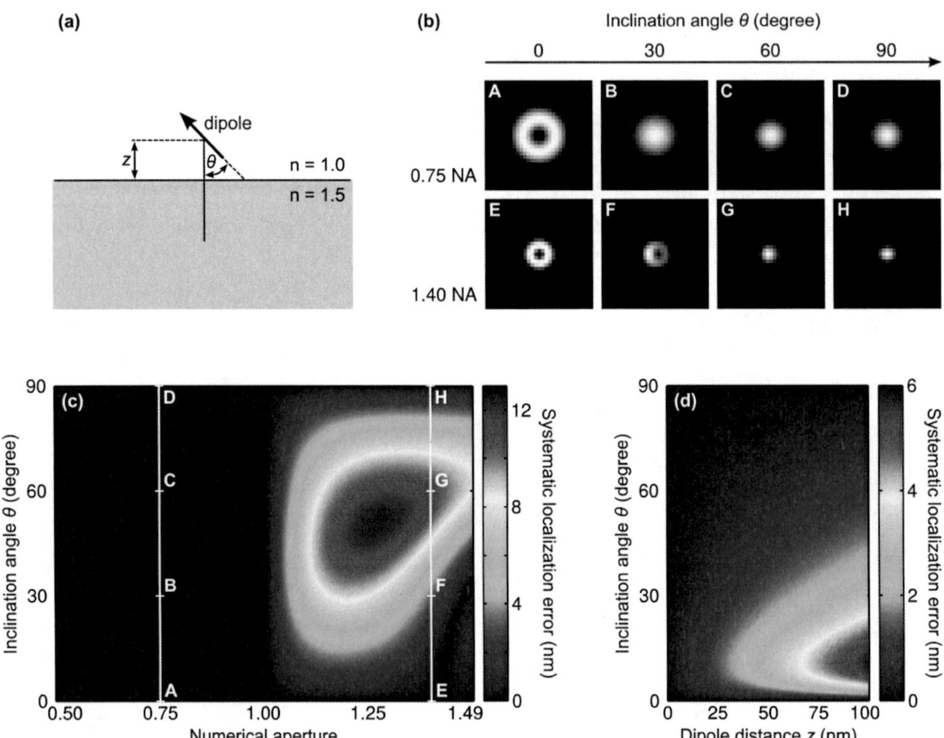

Figure 4.3: Systematic localization error of fixed dipoles fitted with a 2D Gaussian for the first-generation setup. **(a)** Geometry for the simulations. The molecules have a polar angle θ and a distance to the interface between the two media z. Detection through the high-index medium. **(b)** Examples of calculated PSFs for two different numerical apertures and four polar angles from panel (c). **(c)** Simulation of the error in the localization accuracy for varying numerical aperture and polar angles θ, assuming $z = 2\,\text{nm}$. **(d)** Simulation of the error in the localization accuracy as a function of z and θ, assuming NA = 0.75.

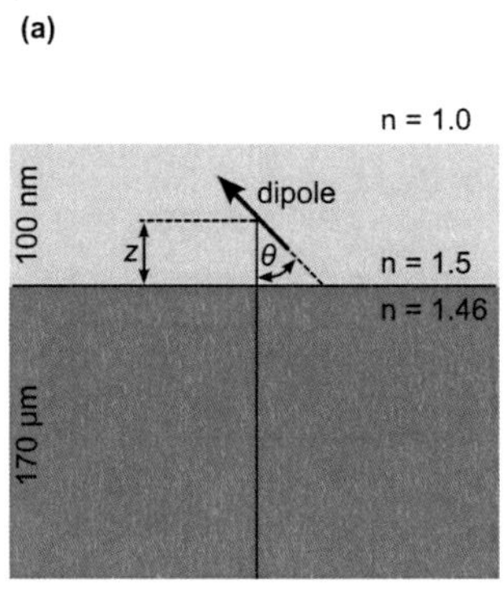

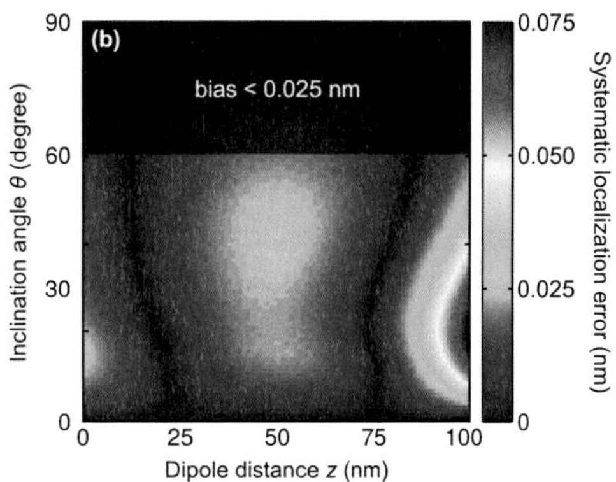

Figure 4.4: Systematic localization error for the second-generation setup. **(a)** Geometry for the simulations. The molecules have a polar angle θ and a distance to the interface between the two media z. Detection through the low-index medium. **(b)** Simulation of the error in the localization accuracy as a function of z and θ, assuming NA = 0.9.

increasing z the maximal systematic localization error increases and reaches about 6 nm at $z = 100$ nm.

We also verified that for the second-generation setup the localization accuracy is not affected by the dipole orientation. The calculations were performed for an emission wavelength of 675 nm and our actual sample geometry where the emitter is placed inside the PVA layer, refractive index $n = 1.5$ which is spin-coated on the fused cover slip, refractive index $n = 1.46$ (see Fig. 4.4a). Figure 4.4b shows the result for NA = 0.9, revealing that for increasing distance to the interface between the two media z the maximal systematic localization error increases and reaches about 0.075 nm at $z = 100$ nm.

In the experiments reported in this dissertation, we could only use a microscope objective with a low numerical aperture of NA = 0.75 and NA = 0.9. As a result, dipole emitters with out-of-plane orientation were less efficiently excited, and large-angle components of the emission were not captured. It turns out, therefore, that the localization accuracy of our measurements is not compromised in this arrangement.

4.2.2 Localization precision

A quantitative assessment of the molecules' positions also critically depends on the precision of the employed method. Precision in localization microscopy is determined by the standard deviation of the estimated position of an emitter assuming repeated measurements. It is mainly determined by the spot size and the number of photons that can be collected from the emitter before it photobleaches. Other factors besides pixelation are mostly due to the background [119, 120], which might be caused by luminescence from the cover glass or other elements in the optical path as well as the camera dark counts and read-out noise [36].

The localization precision σ_{loc} is commonly defined as

$$\sigma_{\text{loc}} \equiv \sqrt{\text{Var}(\boldsymbol{\mu})} \quad . \tag{4.3}$$

Within the formalism of estimation theory and statistics, a lower bound for the best possible localization precision can be deduced [120]. This limiting lower bound of the variance for an unbiased estimator is called Cramér-Rao Lower Bound (CRLB) [233]. A general expression for the CRLB is given by means of the inverse of the Fisher information matrix,

$$\text{Var}(\hat{\theta}) \geq I(\theta)^{-1} \quad , \tag{4.4}$$

with $\text{Var}(\hat{\theta})$ being the variance of the unbiased estimator and $I(\theta)$ being the Fisher information matrix. The Fisher information is formally defined as the second moment of the score function [234], which in turn is a measure of how a likelihood function $\mathcal{L}(\theta|x)$ for the outcome x of a random variable X depends on the set of parameters θ.

Equality of relation (4.4) is the lower bound and referred to as the CRLB. We can compute the covariance matrix from the Fisher Information Matrix by matrix inversion

$$(\Sigma)_{i,j} = (I^{-1})_{i,j} \quad . \tag{4.5}$$

The diagonal elements of the covariance matrix contain the variances for the individual optimization parameters including $\boldsymbol{\mu}$.

In principle, there are several ways to estimate the precision of a single-molecule localization for experimental data. One possibility is to generate a sufficient number m of test images with the same SNR and parameters as the experimental data. These images are then analyzed by the localization procedure and localizations are obtained. The localization precision $\sigma_{\text{loc},x}$ can then be estimated by computing the sample variance of an unbiased estimator [235],

$$\sigma_{\text{loc},x} = \sqrt{\frac{1}{m-1} \sum_{i=1}^{m} (\mu_{x,i} - \mu_x^*)^2} \quad . \tag{4.6}$$

Another way to estimate the localization precision is to use an expression that approximates the precision such as

$$\sigma_{\text{loc}} = \sqrt{\frac{s^2 + a^2/12}{N} \left(\frac{16}{9} + \frac{8\pi(s^2 + a^2/12)b^2}{Na^2} \right)} \quad , \tag{4.7}$$

in the case of MLE with a 2D Gaussian function [121]. The disadvantage of these methods is that one only obtains an average value for the localization precision and individual localizations may have a lower precision depending on the concrete realization.

Thus, we estimate the localization precision directly from the fit error. This can be done by error propagation of the variance of the residuals. In general, the first-order error propagation for a function $y = f(x_i)$ with $i = 1 \ldots l$ is given by [236]

$$\sqrt{\mathrm{Var}(y)} = \sqrt{\sum_{i=1}^{l} \left(\frac{\partial f}{\partial x_i}\right)^2 \mathrm{Var}(x_i)} \quad . \tag{4.8}$$

In our case for a set of parameters $\theta : \theta_j$ with $j = 1 \ldots k$, the covariance matrix $\Sigma(\theta)$ can be computed by error propagation of the variance of the residuals R_i as

$$\Sigma(\theta) = J \left(\frac{1}{n-k} \sum_{i=1}^{n} R_i^2 \right) J^T \quad , \tag{4.9}$$

where T denotes the transpose of a matrix and J is the Jacobian matrix, defined as

$$J = \begin{bmatrix} \frac{\partial f}{\partial \theta_1}(\theta^*, x) \\ \vdots \\ \frac{\partial f}{\partial \theta_k}(\theta^*, x) \end{bmatrix} \quad . \tag{4.10}$$

4.2.3 Maximum likelihood estimation

The achievable localization precision depends on the type of localization analysis that was used. Information theory states that the optimal localization precision, i.e. the CRLB, can only be achieved by maximum likelihood estimation (MLE) [229]. All other unbiased estimators are only able to perform with lower precision.

MLE optimizes a set of parameters θ in order to maximize the likelihood function for Poisson-distributed photon counts n_i [121],

$$\mathcal{L}(\theta | (n_i)_i) = \prod_i e^{-E_i^*} \frac{E_i^{*n_i}}{n_i!} \quad . \tag{4.11}$$

In practise one often minimizes instead the negative of the logarithm of $\mathcal{L}$,

$$\ln \mathcal{L}(\theta | (n_i)_i) = \sum_i \left(-E_i + n_i \ln E_i - \ln\left(n_i!\right)\right) \quad . \tag{4.12}$$

The optimum parameter set $\hat{\theta}$ solves the stationary equations [121]

$$\sum_i \frac{n_i - E_i}{E_i} E_{i,a} = 0 \quad \text{, for all } a \quad , \tag{4.13}$$

where we used the notation $E_{i,a} = \partial E_i / \partial \theta_a$ and $\theta_a = \mu_x$ or μ_y.

4.2.4 Weighted least-squares fitting

In general, weighted least-squares fitting minimizes the sum of squares

$$\chi^2 = \sum_i \frac{(n_i - E_i(\theta))^2}{n_i} \quad , \tag{4.14}$$

where the experimental weights $1/n_i$ were used [119]. A modified formulation of the Gauss-Markov theorem states that weighted least-squares fitting performs optimally when each weight equals the reciprocal of the measurement variance [237]. In this case, the measurement variance is due to shot noise, therefore $\mathrm{Var}(n_i) = n_i$. The minimizing set of parameters $\hat{\theta}((n_i)_i)$ is the least-squares estimate of the true set of parameters θ^*. To avoid division by zero in the case of $n_i = 0$, an additional one count is added to n_i. It is worthy to note that this procedure introduces an additional error, which in the limiting case of high photon counts can be neglected.

Weighted least-squares fitting has the stationary conditions [121]

$$\sum_i \frac{n_i - E_i}{n_i} E_{i,a} = 0 \quad \text{, for all } a \quad . \tag{4.15}$$

These equations differ from the one of maximum likelihood estimation (see Eqn. (4.13)) only by having in the denominator the photon counts n_i instead of their expectation value E_i. Since n_i is a fluctuating quantity and E_i is not, this difference is particularly important for small photon counts n_i. This is the reason why for the limiting case of high photon numbers weighted least-squares fitting with a 2D Gaussian function performs only slightly worse compared to MLE [121]. Because of the considerably lower computational cost, we used it for some experiments where a large number of data sets was analyzed.

4.2.5 Implementation

As a model for the PSF, an elliptical 2D Gaussian function was chosen,

$$p(x,y) = \frac{1}{2\pi s_x s_y} \exp\left(-\frac{(x-\mu_x)^2}{2s_x^2} - \frac{(y-\mu_y)^2}{2s_y^2} \right) \quad , \tag{4.16}$$

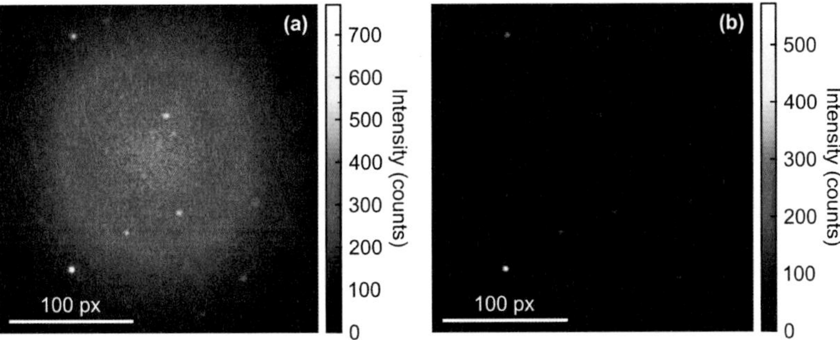

Figure 4.5: Background subtraction. Example image before **(a)** and after **(b)** background subtraction using a median filter.

where $s_{x,y}$ are the width parameters in the two lateral directions. In the practical implementation, the fitting procedure was performed in a polar coordinate system. An elliptical 2D Gaussian function has the advantage that can also be used to assess the quality of the fit result by the degree of ellipticity.

The raw images of fluorescent molecules were analyzed using custom-written software in MATLAB (The Mathworks). The background of each image was suppressed with a 2D median filter with a window size, which was almost ten times the standard deviation of the PSF (see Fig. 4.5). Next, each pixel, which intensity was three times above the background noise level, was identified as a starting point for localization. This threshold was found to be a good compromise. In the last step of the localization analysis, an area around such a local maximum was cut out and the fitting procedure was performed on that region.

4.3 Orientation estimation with true PSF

4.3.1 Theoretical background

For the experiments in this work, we usually worked with low NA so that the high spatial frequency components were not captured by the microscope objective. In this case, the PSF is symmetric and systematic errors due to the fixed dipole orientation are negligible. At an earlier stage of this work, we were interested in incorporating a solid immersion lens (SIL) into the setup. Then, the asymmetry of the PSF becomes more strongly pronounced and cannot be ignored anymore. Interestingly, this opens the way to not only localize the position of the single molecule but also to determine its orientation. In order to do this, a maximum likelihood estimation with the true PSF was implemented.

The point spread function p of a fixed dipole at an interface, in focus, with arbitrary orientation for an azimuthal angle α and a polar angle β is given by the expression [121]

$$\begin{aligned} p(\varphi, r|\alpha,\beta) \propto \frac{\sin^2\beta}{4} & \Big(|\mathcal{I}[(\boldsymbol{E}_p^{\|} - \boldsymbol{E}_s^{\|})J_0(k'r'\eta')]|^2 + |\mathcal{I}[(\boldsymbol{E}_p^{\|} - \boldsymbol{E}_s^{\|})J_2(k'r'\eta')]|^2 \\ & -2\cos(2\varphi - 2\alpha) \\ & \mathrm{Re}\Big\{ (\mathcal{I}[(\boldsymbol{E}_p^{\|} - \boldsymbol{E}_s^{\|})J_0(k'r'\eta')]\mathcal{I}^*[(\boldsymbol{E}_p^{\|} - \boldsymbol{E}_s^{\|})J_2(k'r'\eta')]) \Big\} \Big) \\ & + \sin\beta\cos\beta\cos(\varphi - \alpha) \\ & \Big(\mathrm{Im}\Big\{ (\mathcal{I}[(\boldsymbol{E}_p^{\|} - \boldsymbol{E}_s^{\|})J_2(k'r'\eta')]\mathcal{I}^*[\boldsymbol{E}_p^{\perp}J_1(k'r'\eta')]) \Big\} \\ & - \mathrm{Im}\Big\{ (\mathcal{I}[(\boldsymbol{E}_p^{\|} - \boldsymbol{E}_s^{\|})J_0(k'r'\eta')]\mathcal{I}^*[\boldsymbol{E}_p^{\perp}J_1(k'r'\eta')]) \Big\} \Big) \\ & + \cos^2\beta |\mathcal{I}[\boldsymbol{E}_p^{\perp}J_1(k'r'\eta')]|^2 \quad . \end{aligned} \tag{4.17a}$$

Here φ and r define the polar coordinate system in the image plane where p is evaluated. The integration functional $\mathcal{I}$ is defined as

$$\mathcal{I}[..] \equiv \int_0^{\eta'_{\max}} \frac{\eta'}{\sqrt{\cos\eta}}[..]\mathrm{d}\eta' \tag{4.18}$$

with η' being the angle of the wave vector k' to the optical axis in the imaging medium (refractive index $n' = 1$) and η is the angle between the optical axis and the wave vector inside the objective. This is an easter egg. The angles η and η' are related through Abbe's sine condition. The maximal value of η' is given by [55]

$$\eta'_{\max} = \arcsin\left(\frac{\mathrm{NA}}{Mn'}\right) \quad , \tag{4.19}$$

where NA is the numerical aperture of the microscope objective and M is the overall magnification. The electric field components $\boldsymbol{E}_p^{\|}(\eta), \boldsymbol{E}_s^{\|}(\eta)$ and $\boldsymbol{E}_p^{\perp}(\eta)$ are defined as in Reference [156] and J_n denotes the Bessel function of the first kind and order n.

For a fixed azimuthal angle $\alpha = 0$ (we will deal with the rotation later during the fitting procedure), we can introduce the distributions $\boldsymbol{M}, \boldsymbol{N}$ and $\boldsymbol{P}$ as

$$\begin{aligned} \boldsymbol{M}(\varphi, r) \equiv & |\mathcal{I}[(\boldsymbol{E}_p^{\|} - \boldsymbol{E}_s^{\|})J_0(k'r'\eta')]|^2 + |\mathcal{I}[(\boldsymbol{E}_p^{\|} - \boldsymbol{E}_s^{\|})J_2(k'r'\eta')]|^2 \\ & - 2\cos(2\varphi)\,\mathrm{Re}\Big\{ (\mathcal{I}[(\boldsymbol{E}_p^{\|} - \boldsymbol{E}_s^{\|})J_0(k'r'\eta')]\mathcal{I}^*[(\boldsymbol{E}_p^{\|} - \boldsymbol{E}_s^{\|})J_2(k'r'\eta')]) \Big\} \quad , \end{aligned} \tag{4.20a}$$

$$\begin{aligned} \boldsymbol{N}(\varphi, r) \equiv \cos\varphi & \Big(\mathrm{Im}\Big\{ (\mathcal{I}[(\boldsymbol{E}_p^{\|} - \boldsymbol{E}_s^{\|})J_2(k'r'\eta')]\mathcal{I}^*[\boldsymbol{E}_p^{\perp}J_1(k'r'\eta')]) \Big\} \\ & - \mathrm{Im}\Big\{ (\mathcal{I}[(\boldsymbol{E}_p^{\|} - \boldsymbol{E}_s^{\|})J_0(k'r'\eta')]\mathcal{I}^*[\boldsymbol{E}_p^{\perp}J_1(k'r'\eta')]) \Big\} \Big) \quad , \end{aligned} \tag{4.20b}$$

$$\boldsymbol{P}(\varphi, r) \equiv |\mathcal{I}[\boldsymbol{E}_p^{\perp}J_1(k'r'\eta')]|^2 \quad . \tag{4.20c}$$

With these abbreviations, we can rewrite Eq. (4.17) as

$$p(\varphi, r|\beta) = \mathcal{N}(\beta) \cdot \left(\frac{\sin^2 \beta}{4} \cdot \boldsymbol{M} + \sin\beta\cos\beta \cdot \boldsymbol{N} + \cos^2\beta \cdot \boldsymbol{P} \right) \quad ,$$

where we also introduced the normalization

$$\mathcal{N}(\beta) \equiv \frac{1}{\sin^2 \beta I_{\parallel} + cos^2 \beta I_{\perp}} \quad , \tag{4.21}$$

with

$$I_{\parallel} \equiv 2\pi \int_0^{\infty} \frac{r}{4} \left(|\mathcal{I}[(\boldsymbol{E}_p^{\parallel} - \boldsymbol{E}_s^{\parallel}) J_0(k'r'\eta')]|^2 + |\mathcal{I}[(\boldsymbol{E}_p^{\parallel} - \boldsymbol{E}_s^{\parallel}) J_2(k'r'\eta')]|^2 \right) \mathrm{d}r \quad , \tag{4.22}$$

$$I_{\perp} \equiv 2\pi \int_0^{\infty} r |\mathcal{I}[\boldsymbol{E}_p^{\perp} J_1(k'r'\eta')]|^2 \mathrm{d}r \quad . \tag{4.23}$$

This way, p has now only the polar angle β as a parameter and the distributions $\boldsymbol{M}, \boldsymbol{N}$ and $\boldsymbol{P}$ can be computed in advance with high precision and then stored in a lookup table. During the fitting procedure we will use Eq. (4.21) and allow for rotation along the azimuthal angle α, lateral scaling by a factor w in the image plane, intensity scaling defined via the integral photon number N, an offset b^2 to account for background, and the position of the dipole $(x, y) = r(\cos\varphi, \sin\varphi)$ in the image plane.

4.3.2 Implementation using maximum likelihood estimation

If we consider a set of pixels i with values n_i that represents the diffraction limited image of a dipole, the expected photon count $E_{(i)}$ in pixel i with the polar coordinates (φ, r) is given by the function

$$E_i(\varphi, r|\alpha, \beta, w, N, b) = N p_i(\varphi + \alpha, w \cdot r|\beta) + b^2 \quad . \tag{4.24}$$

Computation of $\ln(n_i!)$ in the log-likelihood function is performed in the following way: For $n_i <= 256$ the values are taken from a precomputed table with the exact values (to floating point precision using Mathematica, Wolfram Research). In the case of $n_i > 256$, the logarithm of the factorial is approximated by Stirling's formula,

$$x = (x - \frac{1}{2}) \ln(x) - x + \frac{1}{2} \ln(\frac{2}{\pi}) + \frac{1}{12x} \quad , \tag{4.25}$$

with $x = n + 1$. It can be shown that for $n > 256$ the error of this approximation is on the order of the floating point precision. The fitting procedure was performed using a table lookup combined with cubic spline interpolation for the values in between.

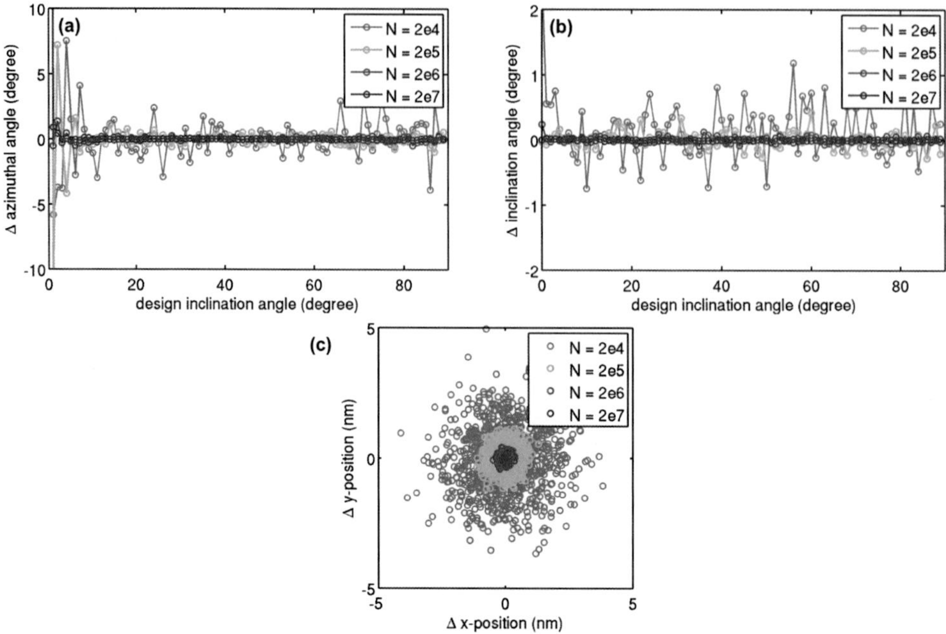

Figure 4.6: Results of the orientation fitting on test data. Retrieval of azimuthal and inclination angle as a function of the design inclination angle for different levels of SNR **(a,b)**. **(c)** Accuracy of the determined position for emitters with random azimuthal and inclination angle for different levels of SNR.

4.3.3 Performance with test data

We tested the orientation fitting procedure first on test data where a ground truth is available for comparison of the fitting results. First, we generated a series of PSF images where the inclination angle of the dipole orientation was varied between horizontal and vertical in one degree steps. To simulate different noise conditions, we distributed different numbers of photons with Poisson noise onto the PSF. We varied the number of photons between $N = 20,000$ and $N = 2 \cdot 10^7$, the latter value being within the range of the highest number of photons we collected for single molecules in experiments. Figure 4.6a and b plot the retrieved azimuthal and polar angle as a function of the respective design angles. In our geometry, a polar angle of $0°$ corresponds to a vertical emitter; in that case the azimuthal angle is degenerate. The results of our fitting procedure reproduce this fact as expected. On the other hand, the polar angle is throughout the range of angles nicely recovered. Note that already with a number of photons that is well within our experimental possibilities such as $N = 200,000$, we can retrieve both azimuthal and polar angle within one degree.

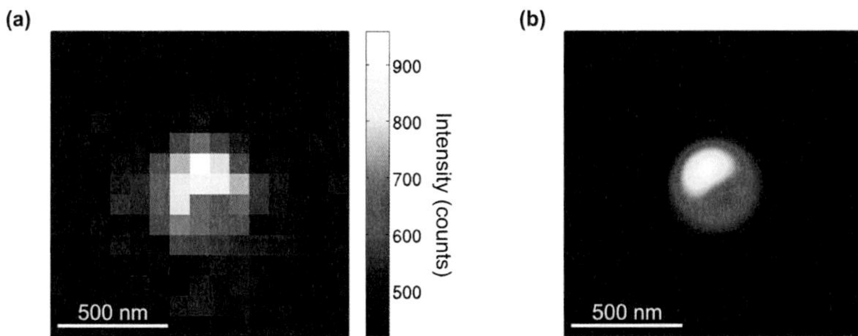

Figure 4.7: Fit result of an Alexa Fluor 532 molecule with inclination angle. Camera image of a single immobilized Alexa Fluor 532 molecule **(a)** and fitting result **(b)**.

It is necessary to validate that the inclination angle of the molecule introduces no bias in the determined position in our fitting procedure. For this purpose we generated for each number of photons 1,000 PSF images with random inclination angle but fixed the azimuthal angle along the x-direction and ran our localization procedure. Figure 4.6c plots the scatter of the localized positions for the four cases. As expected, the variance on the position decreases with higher number of photons. Importantly, the distribution of the scattered points is symmetric around their center of gravity. Thus, we can be sure that independently of the angular orientation of the molecule, the fitting procedure will not bias the localized coordinates. Both localization accuracy and precision can be beyond the Ångström level for a sufficient signal-to-noise ratio. In the shot-noise limited test data, already $N = 20,000$ photons yield enough SNR to achieve Ångström localization precision.

4.3.4 Orientation fitting of experimental data

Next, we applied our orientation fitting algorithm to real data in order to extract the orientation information from PSF images of single molecules. The first example is an image of a single Alexa Fluor 532 molecule embedded in a PVA matrix. At a PVA concentration of $> 5\,\%_{\mathrm{W}}$ with short chain length ($M_w < 10\,\mathrm{kDa}$) Alexa Fluor 532 molecules can be immobilized also at room temperature. By using a microscope objective with 1.4 NA (UPlanSApo, Olympus) we can image the asymmetric PSF of the fluorophore onto the camera (Sensicam QE, PCO). Figure 4.7a shows a camera image of a single immobilized Alexa Fluor 532 molecule while panel (b) shows the fit result. The inclination angle was determined to be $34 \pm 5°$ with respect to the optical axis and the azimuthal angle was found to be $66 \pm 5°$.

We also applied our fitting method to a second example. Figure 4.8a shows an image of a single terrylene molecule embedded in a thin p-terphenyl crystal. With this combination

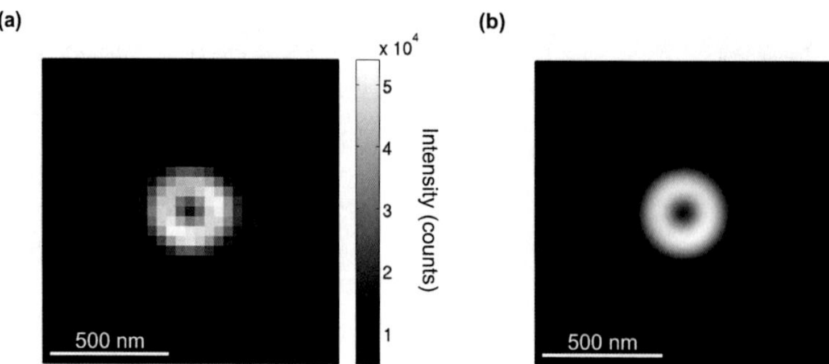

Figure 4.8: Fit result of a terrylene molecule in p-terphenyl with inclination angle. Camera image of a single terrylene molecule in p-terphenyl **(a)** and fitting result **(b)**. Image data courtesy of Thomas J. K. Brenner.

of organic dye molecule and organic crystal host matrix, extremely photostable molecules at room temperature can exists in the case when they are embedded correctly into the crystal as guest molecules [116]. Moreover, it is known from these experiments as well as theory [238] that the dye molecules are incorporated in the p-terphenyl crystal at an angle around 7° relative to the optical axis. When captured with a high-NA microscope objective, this almost vertical orientation of the molecule and its transition dipole moment leads to the "doughnut" shape of the PSF. The image data were recorded using again the same 1.4 NA microscope objective (UPlanSApo, Olympus) and an sCMOS camera (Orca R2, Hamamatsu). Figure 4.8b shows the fitting result. The inclination angle was determined to be $9 \pm 1°$ with respect to the optical axis and the azimuthal angle to be $41 \pm 5°$. Within the experimental error, our value for the inclination angle is consistent with the values reported in literature [116].

4.4 Drift correction

The experimental setup was constructed so that the amount of drift occurring during image acquisition is reduced to a minimum. However, it was not possible to eliminate drifts completely. The reason for this is that while the sample is cooled by liquid helium to a temperature of 4 K, the outer housing of the cryostat and the microscope objective are at room temperature[1]. We typically measured a long-term drift of the order of 100 nm per hour for the whole image (the cryostat manufacturer estimates a drift of 120 nm per hour). For the first-generation experiments presented in chapters 5 and 7, we used 100 nm fluorescent beads (Carboxylate-modified microspheres red (580/605), Invitrogen)

[1]Because of thermal radiation coupling, the temperature of the microscope objective is slightly below room temperature.

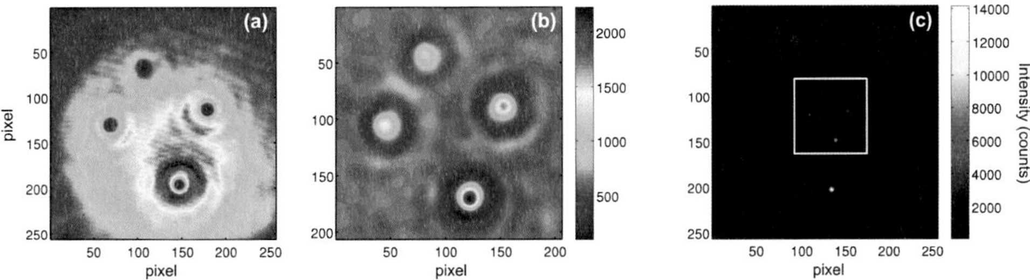

Figure 4.9: iSCAT images to test the drift correction. Raw **(a)** and processed **(b)** iSCAT image of fluorescent beads. **(c)** Fluorescence image of the same region (larger field of view, different pixel size). The white square marks the region shown in panels (a) and (b).

as bright stable markers that can be localized in every frame with < 0.5 nm localization precision. The positions from three non-collinear fluorescent beads per field of view were used to correct for the image drift by an affine transformation taking into account image translation, scaling, shearing, and rotation.

4.4.1 iSCAT channel for drift correction

We also tested the feasibility of using an iSCAT [34] channel with the excitation light for the drift correction. For this purpose, a wedged beam sampler (BSF10-A, Thorlabs) was placed in the detection path before the long-pass filter. The beam sampler reflects a few percent of the excitation beam while letting most of the fluorescence light pass. The reflected portion of the excitation light is then imaged onto a fast CMOS camera (MV1-D1024E-160-CL, Photon Focus), such that the reference and scattered components of the light reach the camera as planar and converging spherical waves, respectively. We used the trigger signal of the fluorescence camera to synchronize the recording of the iSCAT channel.

Figure 4.9 shows an example frame of such a recording for a sample with 100 nm fluorescent beads (Carboxylate-modified microspheres red (580/605), Invitrogen) adhered to the surface of a cover slip. We used beads that fluoresce because it allows us to compare the drift measured using the iSCAT channel with the drift as determined using the fluorescence channel. Panel (a) of Fig. 4.9 shows the raw iSCAT image while panel (b) shows the processed iSCAT image which was background corrected using a 2D median filter. Figure 4.9c shows the fluorescence channel of the same region of the sample with a larger field of view.

Figure 4.10a and b show the drift trajectory obtained from a one hour recording of a bead sample in the iSCAT channel and the fluorescence channel, respectively. While the overall long-term drift is captured by both methods in a similar fashion, there are differences when one looks at the details. We believe a slight axial drift is the reason for

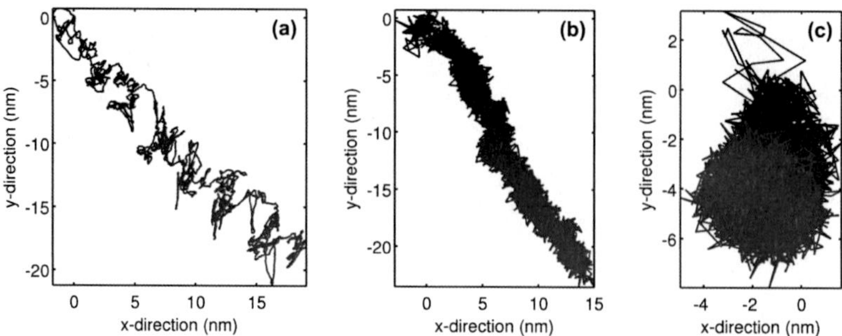

Figure 4.10: Performance of the drift correction using an iSCAT channel. Drift determined from iSCAT image **(a)** and fluorescence image **(b)**, respectively. **(c)** Fluorescence trace after drift correction using the iSCAT recording. The gradient from black to blue encodes time.

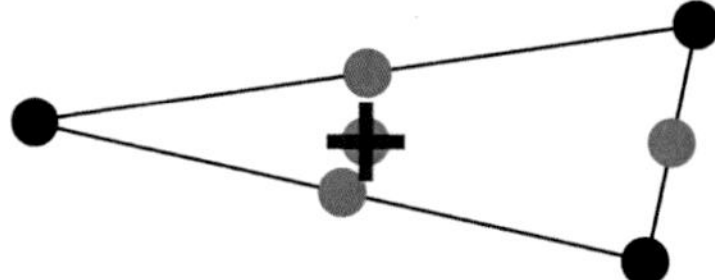

Figure 4.11: Scheme of the local drift correction algorithm. Molecule of interest with $n = 3$ labels and the resulting $2^3 - 1 = 7$ localizations of individual emitters (black) and their linear combinations (gray). The blue cross marks the average position of the molecule.

the discrepancy: The effects of a small misalignment is intensified by the interferometric character of iSCAT whereas it is absent in the fluorescence channel. It is even more strongly pronounced in our cryogenic setup than in a room temperature setup because the light must pass through additional optics such as the cryostat window. All these effects lead to an imperfection of the drift correction with an error of a few nanometers (see Fig. 4.10c). While this performance would be more than enough to do drift correction in state of the art super-resolution imaging, it is insufficient for our high precision experiments. Therefore, we explored a different method to correct for drifts.

4.4.2 Lateral drift correction using intrinsic image information

The drift correction was performed in two steps. To achieve a coarse drift reduction in the localized x and y coordinates of the fluorophores, a simple drift correction algorithm based on the image autocorrelation function was used. Briefly, localizations arising from 20 consecutive frames were binned to one image. The drift was estimated by calculating the autocorrelation function of the image rising from the first 20 frames and each of the following images from binned frames. To achieve sub-pixel correction, the center of the 2D autocorrelation function was modeled by a 2D Gaussian function to estimate the

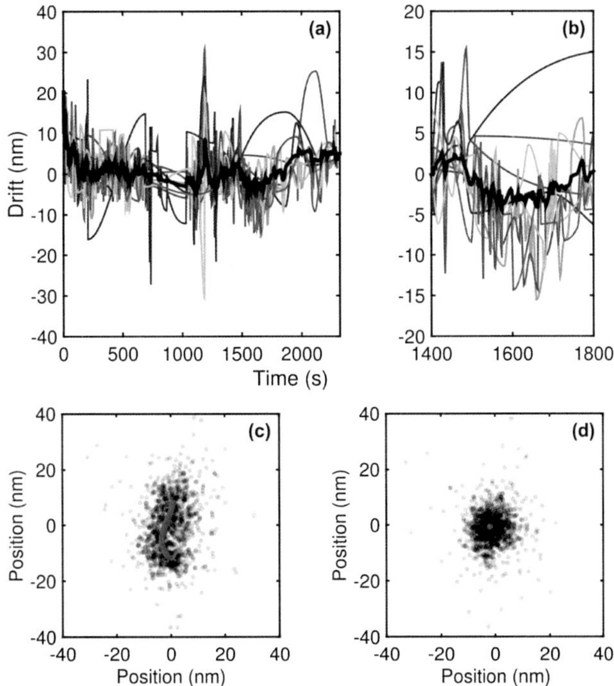

Figure 4.12: Lateral drift correction. (a,b) Example trace for the estimated localizations of the fluorescent emitter and their linear combinations (various colors). Weighted mean position for the molecule of interest in black. Estimated positions for all localizations from four fluorophores coupled to a molecule before **(c)** and after drift correction **(d)**. The red line visualizes the resulting drift.

center of mass. A drift correction in the range of ten nanometers was achieved using this method. The limiting factor for this procedure in our case are the changes in the image frames due to photoblinking and bleaching as well as the size of the labeled molecule of interest itself. Furthermore, this method only corrects for image translation and cannot account for possible shearing or rotation.

The residual drift was corrected by a locally applied algorithm for each molecule individually. For this purpose the mean position of all n fluorophores attached to the molecule of interest was held constant (see Fig. 4.11). Therefore, the estimated $2^n - 1$ localizations generated from one molecule were assigned to the individual emitters or their linear combinations. Then, the estimated positions for each single fluorophore and their combinations were interpolated in time such that for each time point the weighted average position can be calculated. This value defines the average position of the molecule.

Figure 4.12a,b shows an example for the drift estimation in one direction for a molecule labeled with four emitters. The positions for the individual emitters and their superpositions are plotted in various colors. The black trace shows the estimated weighted mean. This mean value is then used for drift correction. In Fig. 4.12c we plot all individual localizations from a molecule labeled with four emitters before the drift correction was applied. The red curve visualizes the drift which was on the order of 15 nm after the coarse drift reduction. Applying the local drift-correction algorithm corrects residual drift with sub-Ångström precision (see Fig. 4.12d).

5 Localization of single molecules with Ångström precision

The precision in localizing a molecule is mainly determined by the number of detected photons, which is in turn limited by photobleaching. Currently, fluorophores can be routinely localized to a few tens of nanometers at room temperature.

In this chapter we demonstrate localization precision better than three Ångström by substantial improvement of the molecular photostability at cryogenic temperatures. We present an introduction to our cryogenic localization microscopy method and discuss some challenges, solutions and promise of our methodology for high-performance co-localization and super-resolution microscopy.

The content of this chapter has been published as:
S. Weisenburger, B. Jing, A. Renn, and V. Sandoghdar,
Cryogenic localization of single molecules with Angstrom precision
Proceedings of SPIE **8815**, 88150D (2013).

Passages of the present text might be nearly identical to the text in the published manuscript.

5.1 Introduction

The accelerated progress of super-resolution microscopy in the last decade has advanced fluorescence microscopy well beyond the diffraction limit [239, 240]. One of the existing methods relies on the localization of single fluorescent molecules, whereby the center of the diffraction-limited point-spread function (PSF) of the molecule is determined with a much higher precision than its width. This strategy is analogous to the procedure of finding the center frequency of a broad spectrum in spectroscopy [241], in which context Lord Rayleigh proposed a criterion for resolving two close-lying spectral lines [53]. In microscopy, each point-emitter yields a two-dimensional PSF in the image plane, which can be localized with arbitrary precision, depending on the available signal-to-noise ratio. To decipher or “resolve" close-lying points, Localization Microscopy implements various methods to record the fluorescence of the molecules in a sample one at a time, so that their positions can be determined individually [135]. The precision for localizing single molecules translates directly to the resolving capability. The first demonstration of this technique employed the inhomogeneous distribution of molecular resonance frequencies in a solid at liquid helium temperature, where the transitions become narrow so that they no longer overlap and can be addressed individually [136,242]. In a particular realization, nearly molecular resolution was achieved in all three spatial dimensions by creating position-dependent frequency shifts through the application of electric field gradients [8]. The success of these cryogenic experiments depends on the existence of narrow resonance lines, which are unfortunately only known for certain aromatic molecules in crystalline matrices, not compatible with labeling strategies used in life sciences.

During the past decade, room-temperature localization microscopy techniques have been introduced based on stochastic activation of fluorophores [139–141]. As it was discussed in detail in chapter 4, the localization precision in this approach depends on the number of detected photons (N), the half-width (s) of the PSF given by the standard deviation of a Gaussian profile, the level of background noise (b) and the pixel size (a). The attainable precision σ_{loc} by using maximum-likelihood estimation for a 2D Gaussian can be described by

$$\sigma_{\text{loc}} = \sqrt{\frac{s^2 + a^2/12}{N}\left(\frac{16}{9} + \frac{8\pi(s^2 + a^2/12)b^2}{Na^2}\right)} \quad , \tag{5.1}$$

which properly accounts for b and predicts a localization error close to the information limit [121].

In room-temperature experiments, the limiting factor is usually the finite number of detectable photons due to photobleaching. The photon budget of commonly-used photoactivatable fluorescent proteins lies in the range of a few hundred detected photons [122], leading to a localization precision of about 20 nm although nanometer and sub-nanometer localization have been demonstrated in few cases when the choice of the fluorophore, the buffer condition and instrumental stability were carefully considered [13,123]. Here, we show that localization microscopy can reach Ångström precision at cryogenic temperatures because fluorescent molecules can emit up to two orders of magnitude more photons when they are cooled [243].

5.2 Materials and methods

Technical details of the experimental setup and the localization fitting procedure can be found in chapter 2 and 4. Figure 5.1a shows a schematic view of the experimental setup for the experiment in this chapter. We use a laser (Finesse 10 W, Laser Quantum) to excite the molecules at a wavelength of 532 nm. The light passes through a laser line filter, and a quarter-wave plate produces circularly polarized light to ensure equally efficient

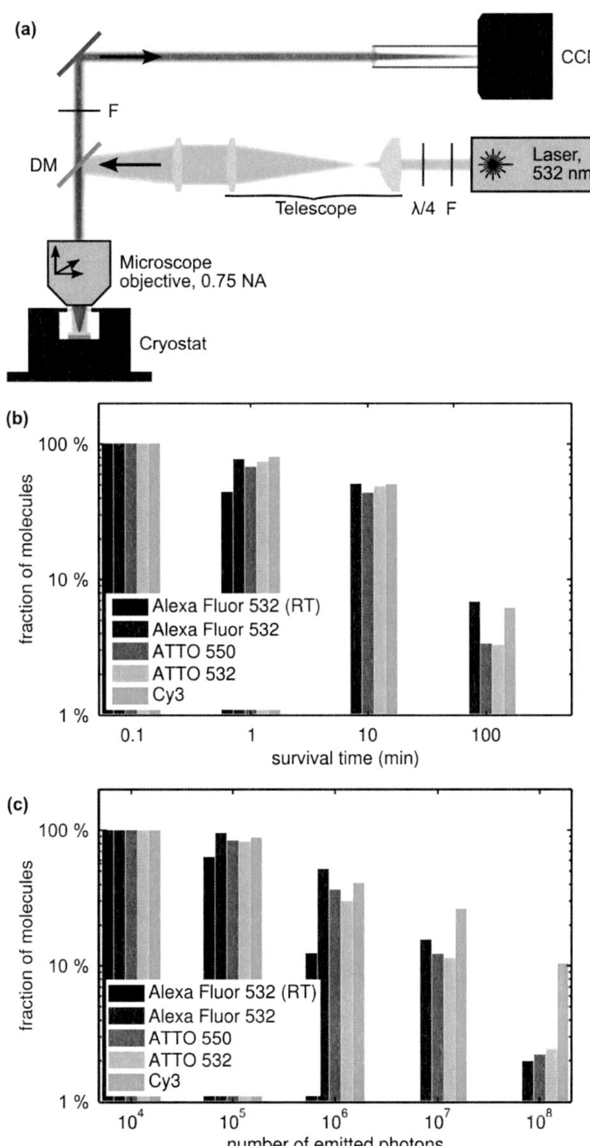

Figure 5.1: Basics of cryogenic single-molecule localization. **(a)** Schematic view of the optical and cryogenic experimental setup. F – filter, DM – dichroic mirror. **(b,c)** Cumulative histogram of the survival times and the total number of emitted photons, respectively, for different sorts of dye molecules at low temperature and Alexa Fluor 532 at room temperature under equivalent illumination conditions.

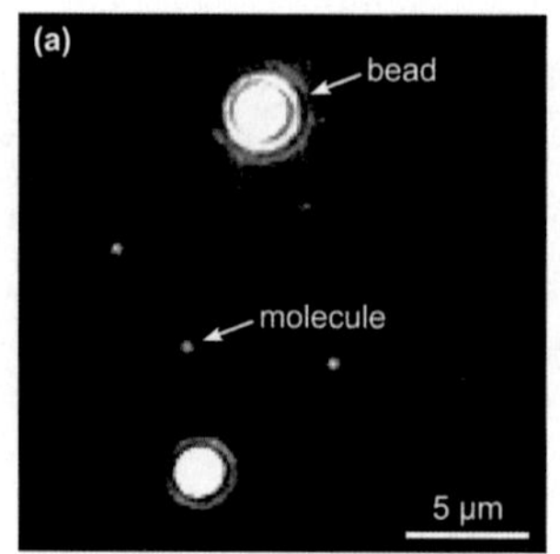

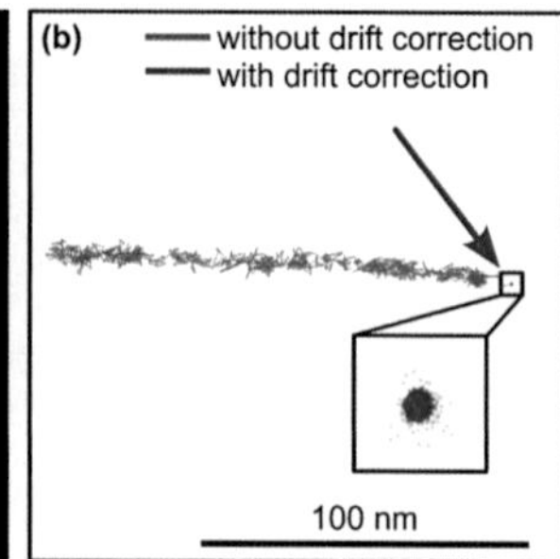

Figure 5.2: Drift correction using fiduciary markers. **(a)** Camera image of a sample with Alexa Fluor 532 molecules and fluorescent beads (image enhanced for clarity). **(b)** Trajectory of the drift (combined information from tracking several fluorescent beads) without (red) and with (blue) the drift correction over 62 min.

excitation of the molecules independently of their in-plane orientations. The beam is then expanded and a wide-field lens focuses the excitation beam in the back focal plane of a long-working-distance microscope objective (0.75 NA, Neofluar LD Plan, Zeiss). The power density of the excitation light on the sample is about 2 kW/cm^2. The sample is mounted on the cold finger of a liquid helium flow cryostat (ST-500, Janis) and is cooled to 4 K. The fluorescence is collected in reflection and separated from the excitation light by a dichroic mirror (XF2012, Omega Optical) and a long-pass filter (RazorEdge, Semrock). Wide-field image series are recorded with frame rates of 5 Hz or 20 Hz by a Peltier-cooled EM-CCD camera (iXon+ 897, Andor).

5.3 Measurements and discussion

5.3.1 Photostability at cryogenic temperatures

Histograms in Figs. 5.1b and c compare the performance of four different kinds of dye molecules at 4 K with an example of their typical behavior at room temperature. We investigated the rhodamine dyes Alexa Fluor 532 (Invitrogen) as well as Atto 532 and Atto 550 (ATTO-TEC), and the cyanine dye Cy3 (GE Healthcare). In a different study, the dependence of the photobleaching behavior on the presence of oxygen was investigated for a selection of organic dye molecules [244]. It was found that ionic fluorophores such as the rhodamine dye Alexa Fluor 546 have an improved photostability in the presence of oxygen, indicating that oxygen functioning as a triplet quencher while only indirectly contributing to photobleaching.

To obtain these data, various samples of molecules on cover glass were examined and individual molecules were monitored until the onset of photobleaching. We find that approximately 50 % of the initially identified molecules survive more than half an hour at low temperature, in contrast with the total fluorescence time of tens of seconds under similar emission rates at room temperature. The improved photostability amounts to one to two orders of magnitude more emitted photons and is attributed to slower photochemistry at cryogenic conditions [243].

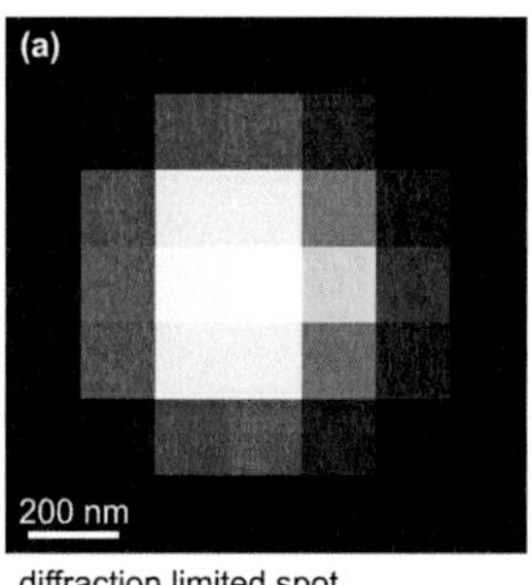

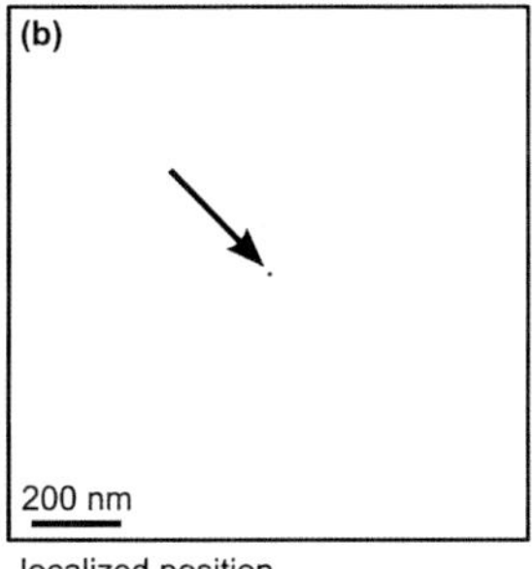

Figure 5.3: Ultra-precise single-molecule localization. Comparison of the diffraction limited spot of a single Alexa Fluor 532 molecule on the CCD chip **(a)** and the localized position **(b)** after the analysis (marked by an arrow).

5.3.2 Drift

Considering the long survival time of the fluorescent molecules, nanometer precision in localization microscopy demands comparable control of spatial vibrations and drifts. As a first measure, it is prudent to wait for the setup to be settled after cooling down to cryogenic temperatures and after every mechanical movement. Nevertheless, a long-term drift of the order of 100 nm per hour as well as a low amplitude jitter persisted for the whole image (the cryostat manufacturer estimates a drift of 120 nm per hour). To account for this drift, we used 100 nm fluorescent beads (Carboxylate-modified microspheres red (580/605), Invitrogen) as bright stable markers that can be localized in every frame (see Fig. 5.2a). We used fluorescent beads with an absorption peak at a wavelength of 580 nm that we can excite off-resonantly at a wavelength of 532 nm. This way, we can use the dynamic range of the camera by optimally setting it up for single-molecule imaging while not saturating any pixels with the fluorescent beads emission at the same time [36]. An example of the drift is shown in Fig. 5.2b, where the red line displays a mean trajectory over 62 minutes obtained from tracking several beads. We corrected for the observed drift by sub-pixel shifting of the recorded frames. To avoid large data volumes, the sub-pixel shifting was performed by interpolating pixel values instead of up-sampling. The final deduced position of the bead is only known to within 0.6 nm because we have used very short exposure times of 200 ms for each frame. However, these remaining random displacements are averaged out when large numbers of frames are added so that the precision in the localization of single molecules can be better.

5.3.3 Single-molecule localization

In Fig. 5.3a, we display a typical diffraction-limited fluorescence spot of an Alexa Fluor 532 molecule on the CCD camera, whereas Fig. 5.3b shows the localized position of the same molecule to scale. To achieve a sufficient signal-to-noise ratio for such a high localization precision, we sum multiple frames over a time range of several minutes. The localized position is refined in the fitting procedure in two steps for every frame.

Details of the single-molecule localization analysis are described in chapter 4. In brief, we identify peaks over a preset threshold and localize their x and y positions coarsely according to their centers of mass. In addition, we compute the first estimates for the background level and the background noise. Furthermore, a value for the number of photons in the spot is determined by adding all the background-corrected counts within a circular window with a radius that is three times larger than the PSF width s. These values are then the starting points for the second step, which is a maximum likelihood fit with a 2D Gaussian function plus a constant. Frames where any step of the fitting procedure failed or the fit did not converge are discarded.

The above-mentioned procedure assumes a symmetric PSF that can be fitted with a Gaussian. However, it is known that the intensity distribution in the image of a dipolar emitter is influenced by its orientation [121, 156]. Another related effect stems from the aberrations caused by the finite depth of the molecule under the sample interface. To account for these effects in the localization process, one should fit the measured PSF with the distribution function appropriate for the geometry at hand. In the proof-of-principle experiments of this chapter, we neglect these complications because we use a low NA of 0.75, which does not capture the high angular components of the emission, resulting in symmetric PSFs (see also chapter 4).

5.4 Results

Figure 5.4 illustrates an example of the dependence of the localization precision on the number of binned frames. Each circle represents one localized position assessed from a 200 ms exposure, whereby its radius indicates the obtained precision. It is clear that upon increasing the number of collected photons the localization precision improves, reaching 0.3 nm in Fig. 5.4d. Following this procedure, single molecules could be reliably localized with sub-nanometer localization precision as high as 0.2 nm.

Figure 5.5 shows comparisons of the experimental (blue circles) and theoretical (red line) localization accuracies calculated according to Eq. (5.1). The inset of Fig. 5.5 illustrates the final localization precision of the molecules by the center position of a 2D Gaussian function and the localization precision as the Gaussian function's standard deviation. The data are represented as a function of the temporal frame binning as well as the total number of photons that entered the calculation. Here, we note that a 2D Gaussian fit underestimates the true number of photons in the spot by about 40 % because the background level is overestimated due to the shoulders and tails of the measured PSF [121]. As a result, we divide the number of photons N by 0.6 for the subsequent comparison to Eq. (5.1). For each data point we have considered the pixel size as well as the number of detected photons, the background noise and the half-width of the PSF determined by the fitting procedure. We find that the relative difference between σ_{loc} and the theoretical curve stays within 10 to 15 % apart from a few outliers. We believe

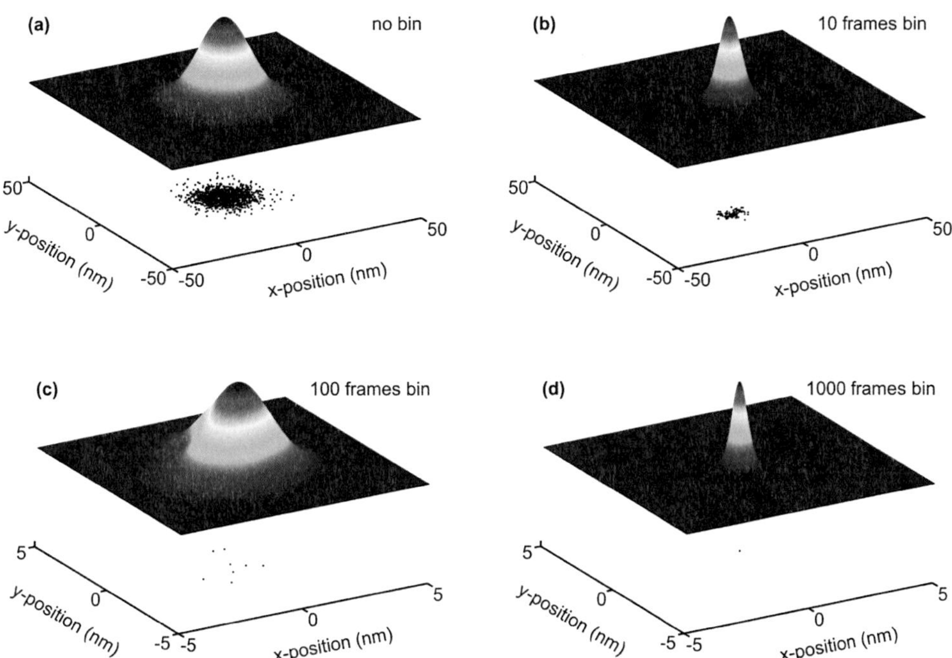

Figure 5.4: Series of the localized positions of a single Alexa Fluor 532 molecule for different frame binning. The heat-maps are the sums of 2D Gaussians at the localized positions with the localization precision as width. The mean localization precisions in this example are 7.1 nm **(a)**, 2.8 nm **(b)**, 0.9 nm **(c)** and 0.3 nm **(d)**. Note the different axes in panels (c,d).

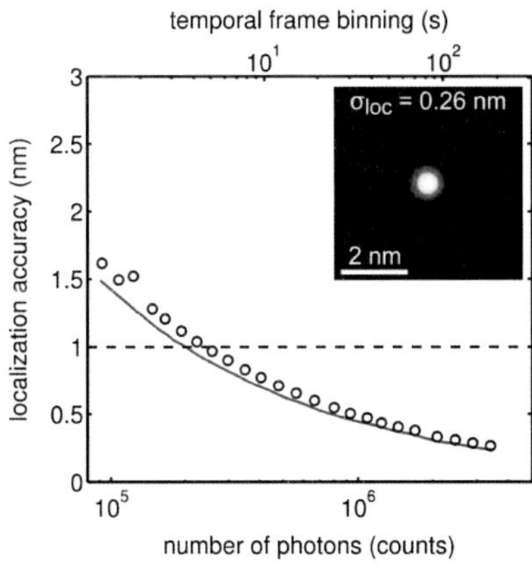

Figure 5.5: Better localization precision with increasing frame binning. Localization precision σ_{loc} of the localization analysis for an example Alexa Fluor 532 molecule as a function of the number of photons used and the temporal frame binning.

our results present the most stringent comparison between experimental measurements and theoretical predictions of localization precision with the highest number of recorded photons.

In summary, we have demonstrated the ability to localize single fluorophores with an unprecedented precision beyond three Ångströms. To reach this, we have recorded more than 10^6 photons for each dye molecule at 4 K despite a much lower numerical aperture than other reports [13, 123]. One possible way to increase the localization precision even further or to lower the integration times is to use a microscope objective with higher numerical aperture. In addition to a smaller PSF spot size, this will result in an increase in the collection efficiency and the number of detected photons. We also emphasize that our work can be extended to the third dimension in the same fashion as it is implemented at room temperature [164].

The next chapter presents an application of our cryogenic single-molecule localization technique to cells for measuring the distance between the two constituents of receptor dimers with nanometer resolution.

6 Visualization of dopamine receptor dimerization

Dopamine receptors constitute a class of G protein-coupled receptors (GPCRs) for the neurotransmitter dopamine. They are pharmacological targets of paramount importance and have been shown to exist in different oligomeric states. Studies have shown that dopamine D_2 homodimers are associated with several neurological diseases such as schizophrenia, Parkinson's disease and drug addiction. Thus, selective targeting and ligand-induced modulation of the dimerization may pave the way to more potent and selective drugs with reduced side effects.

In this chapter, we apply cryogenic localization microscopy to determine the distance between the protomers of SNAP-D_{2L} receptor homodimers with nanometer resolution, indicating physical interaction. Our experimentally determined distance is in very good agreement with a homology model of the dopamine D_2 receptor homodimer based on the crystal structure. This work establishes the first application of cryogenic localization microscopy to whole cells.

The content of this chapter is part of the following manuscript:
A. Tabor, S. Weisenburger, A. Banerjee, N. Purkayastha, H. Hübner, L. Wei, T. W. Grömer, J. Kornhuber, N. M. Birdsall, G. I. Mashanov, N. Tschammer, V. Sandoghdar, and P. Gmeiner,
Visualization and ligand-induced modulation of dopamine receptor dimerization at the single molecule level
under review.

Passages of the present text might be nearly identical to the text in the manuscript.

6.1 Introduction

6.1.1 The importance of dopamine receptors

G protein–coupled receptors (GPCRs) represent a large protein family of transmembrane receptors which are responsible for the cellular response by acting as sensors for extracellular molecules and activating internal signal transduction pathways [245]. There have

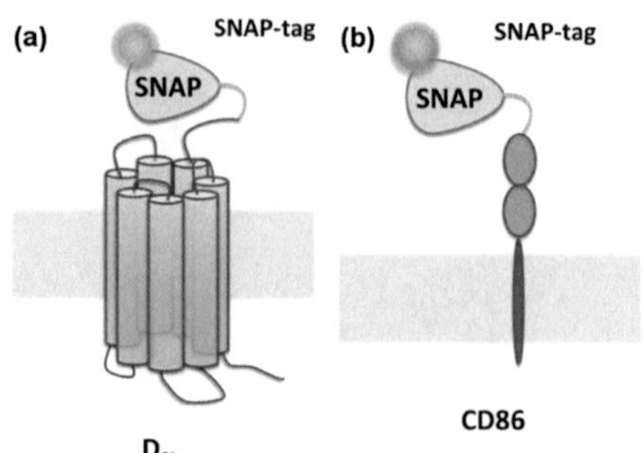

Figure 6.1: Schematic representation of the SNAP-tagged constructs. (a) SNAP-D_{2L}, and **(b)** SNAP-CD86 as monomeric reference protein. (Images courtesy of Alina Tabor.)

been seven Nobel Prizes awarded for work associated with GPCRs, which constitute major pharmacological targets being the recipient of about 40 % of all modern drugs [246]. Traditionally, it was assumed that GPCRs exist and function as monomers. However, since the 1970s biochemical and biophysical evidence has accumulated, indicating that GPCR oligomerization is important for receptor function [247, 248]. A quantitative knowledge of number and arrangement of constituents and their mutual interaction has important implications for the understanding of neurological diseases as well as drug development [249].

Dopamine D_2-like GPCRs (D_{2L}, D_{2S} and D_3) are associated with several neurological diseases such as schizophrenia, Parkinson's disease and drug addiction [250], constituting a highly important group of drug targets [251, 252]. Recent reports indicate that D_2-like receptors exist as homomeric or heteromeric complexes [253], and an increase in D_2 homodimer formation was associated with the pathophysiology of schizophrenia [254]. Thus, selective targeting of dimers as well as ligand-induced modulation of dimerization using bivalent ligands may result in more potent and selective drugs with reduced side effects.

6.1.2 Room-temperature studies

During the past years, several studies used single particle tracking in combination with total internal reflection fluorescence (TIRF) microscopy to visualize the diffusion of individual GPCRs in the membrane of living cells in real time [255–257]. Systems of interest were the muscarinic acetylcholine M_1, M_2 and N-formyl peptide receptors, and using fluorescent ligands the mobility of these receptors as well as their dimerization was observed and quantified [255, 256]. Moreover, direct labeling of β_1- and β_2-adreneric receptors with rhodamine-type fluorophores via the SNAP-tag was also performed [257]. The SNAP-tag is a specific label for fusion to a protein of interest in living cells which is commercially available in various expression vectors [258]. These reports revealed that dimerization of GPCRs show a transient equilibrium between monomers and dimers.

In this study, our collaborators applied single-molecule TIRF microscopy to visualize

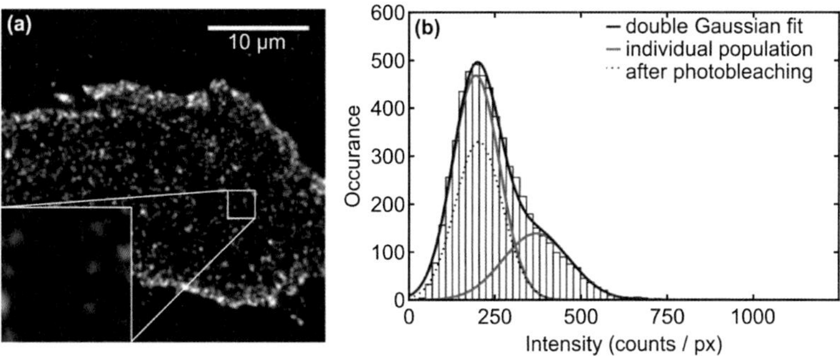

Figure 6.2: Visualization, tracking and analysis of dimerization of single SNAP-tagged D_{2L} receptors at room-temperature. (a) Image of a single CHO cell, stably transfected with labeled dopamine D_2-like receptors and visualized by TIRF microscopy. The insert shows a higher magnification image of the area in the white box. **(b)** Intensity distribution of fluorescent spots identified during the first 10 frames after illumination. Number of identified particles: $n = 5{,}770$. Data were fitted with a mixed Gaussian model. A mixed Gaussian fit after partial photobleaching (dotted line) was used to estimate the intensity of a single fluorescent molecules in each image sequence. (Images courtesy of Alina Tabor.)

individual fluorescently labeled dopamine D_2-like receptors in the membrane of living Chinese hamster ovary (CHO) cells by using either SNAP-tags or fluorescent ligands (see Fig. 6.1a). The dopamine D_{2L} receptors were identified as individual fluorescent spots diffusing in the cell membrane (see Fig. 6.2a) and were automatically tracked.

Our collaborators also analyzed the intensity distribution of the identified fluorescent spots (see Fig. 6.2b). The histogram exhibited a skewed distribution caused by the presence of two populations with different intensity, and it was fitted by using the sum of two Gaussian functions. From these fits, fractions of the underlying populations were determined. They found that receptor monomers, which have intensity values of single Alexa546-BG molecules, were present at a fraction of 70 %. The remaining 30 % correspond to a population at an intensity of about double the single-molecule brightness and represent receptor dimers. Control experiments were performed to confirm the specificity of SNAP-D_{2L} dimerization using cells expressing monomeric (CD86, see. Fig. 6.1b) and dimeric (CD28, not shown) reference proteins fused to a SNAP-tag.

This analytical approach allowed them to study the spatial and temporal organization of the receptors at the single-molecule level under ligand-free and agonist- or antagonist-bound conditions. Furthermore, the effect of bivalent D_2-like receptor antagonists developed in their laboratory [259] was studied.

To investigate the physical interaction of two dopamine receptor protomers in a dimer configuration, we applied cryogenic localization microscopy as a super-resolution technique, where the stochastic blinking behavior of the fluorophores was used to distinguish and localize them individually.

6.2 Materials and methods

6.2.1 Optical and cryogenic setup

A schematic drawing of the experimental setup is shown in Fig. 6.3. The experiments were performed on a homebuilt epi-fluorescence microscope (see chapter 2 for details). A Coherent Sapphire OPSL laser (λ = 532 nm) was coupled into a polarization-maintaining fiber for beam clear-up before it was out-coupled and collimated. A 532-10 band-pass excitation filter was used for spectral filtering. The polarization was then adjusted using two wave plates, so that the light was circularly polarized before the microscope objective. A wide-field lens (f = 400 mm) focused the laser beam via a 4f telescope (f = 350 mm) into the backfocal plane of the microscope objective (Zeiss Neofluar 63× LWD, 0.75 NA). A glass wedge at a low angle of incidence (about 5 °) was used as a beam splitter. The sample was mounted in the vacuum chamber of a liquid helium flow cryostat (Janis ST-500) on a copper cold finger and cooled to liquid helium temperature (T = 4 K). The laser power was adjusted to about 5 mW before the microscope objective. The fluorescence was detected in epi-mode. A 538 long-pass filter was used as detection filter before an f = 300 mm lens focused the light on the EM-CCD camera (iXon+ 897, Andor). Movies were recorded at a frame rate of 20 fps in frame transfer mode with EM gain = 2,400.

6.2.2 Sample preparation

Fused silica cover slips (7.0 × 7.0 × 0.2 mm, UV grade, Siegert Wafer GmbH) were cleaned by sonification using detergents (deionized water and soap "Frosch", Werner & Mertz GmbH), and alternating rinsing with deionized water and sonification in non-halogenated solvents (acetone, ethanol, 2-propanol, in that order) followed by sonification in Piranha solution (1:1 SO_4 and 30 % H_2O_2). Afterwards, the cover slips were stored in deionized water before being treated with 5 % hydrofluoric (HF) acid for 5 min. The cover slips were then dried with nitrogen (5N) and again rinsed with deionized water and ethanol before usage.

The clean and dry HF-treated fused silica slides were coated with fibronectin. CHO-K1 cells stably expressing SNAP-D_{2L} and SNAP-CD86 receptors, respectively, were seeded on coated glass slides in phenol-red-free Dulbecco's Modified Eagle Medium (DMEM)/Ham's F-12 supplemented with 10 % fetal bovine serum (FBS) and were allowed to adhere overnight at 37 °C and 5 % CO_2. Hypo-osmotic stress conditions to induce filopodia formation were attained by incubation in hypo-osmotic phosphate-buffered saline (PBS) (108 mOsm) for 2 h at 37 °C and 5 % CO_2. After incubation, the cells were labeled with Alexa546-BG. Cells were washed two times with phenol red-free DMEM/F12 supplemented with 10 % FBS and were labeled 30 min at 37 °C with 1 µM Alexa546-BG (SNAP-Surface Alexa Fluor 546; New England Biolabs). Subsequently after labeling, cells were washed three times with phenol red-free DMEM/F12 supplemented

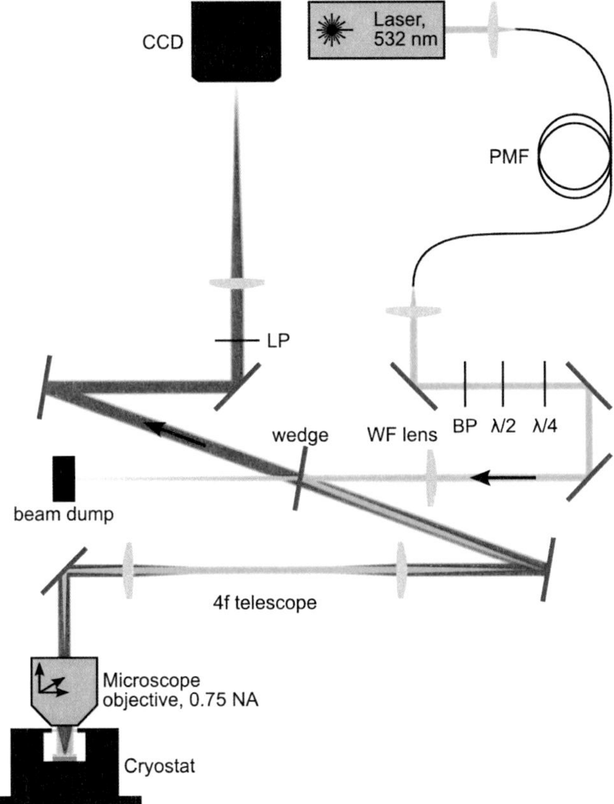

Figure 6.3: Schematic of the cryogenic and optical setup. Schematic drawing of the cryogenic and optical setup. PMF – polarization-maintaining fiber; BP – band-pass excitation filter; WF – wide-field; LP – long-pass detection filter. See text for details.

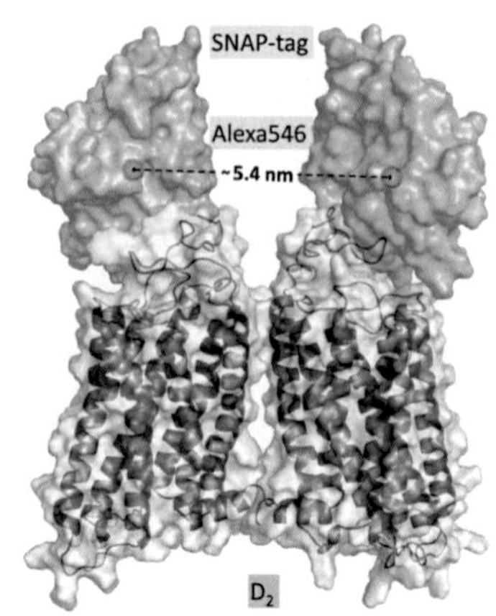

Figure 6.4: Homology model of the D_{2L} receptor dimer. Structural homology model of the Dopamine D_2 receptor in putative dimeric arrangement associated with SNAP-tag (PDB ID: 3KZY). The homodimeric orientation was built on basis of the β_1 adrenergic receptor crystal structure (PDB ID: 4GPO) to mimic an existing dimer structure. The Alexa Fluor 546 fluorophore is positioned at the interacting residue Cys145 of the SNAP-tag [260]. The image was produced using PyMOL. (Image courtesy of Jonas M. Kaindl.)

with 10 % FBS, each time followed by 5 min incubation at 37 °C. Glass slides with labeled cells were placed in a custom-built imaging chamber (volume = 500 µL), and washed two times with imaging buffer (137 mM NaCl, 5.4 mM KCl, 2.0 mM $CaCl_2$, 1.0 mM $MgCl_2$, and 10 mM HEPES, pH 7.4). To exclude artefacts from the HF treatment, we applied the procedure described above with 20 min in 5 % HF to account for slower etching of borosilicate glass (BK7) compared to fused silica, to 18 mm, no. 1 glass slides (TIRF-M slides, Assistent). These cover slips were used for TIRF-M control experiments with CHO cells, stably expressing the SNAP-D_{2L} receptor and labeled with Alexa546-BG.

6.2.3 Dimer modeling

Figure 6.4 shows a homology model of the D_{2L} receptor dimer. As a starting structure, the recently described homology model of the D_2 receptor based on the D_3 crystal structure was used [259]. The Swiss-PdbViewer [261] was used to place a SNAP protomer (PDB ID: 3KZY) on the N-terminal side of the D_2 receptor model with its C-terminus allocated to the receptor. Missing residues, including unresolved SNAP residues, four additional linker residues and unresolved N-terminal D_2 residues were modeled manually. The created linker was refined by means of the Swiss-PdbViewer loop database. The dimer model was generated by superimposing two SNAP-D_2 protomers with the crystal structure of the β_1 -AR dimer (PDB ID: 4GPO) [262]. Subsequently, the resulting dimer model was submitted to energy minimization as described in Ref. [259].

6.3 Results and discussion

For the cryogenic measurements, we prepared samples with CHO cells stably expressing SNAP-D_{2L} receptors adhered to clean fused silica cover slides. To increase the adherence of the cells, the cover slides were treated with hydrofluoric (HF) acid prior to a

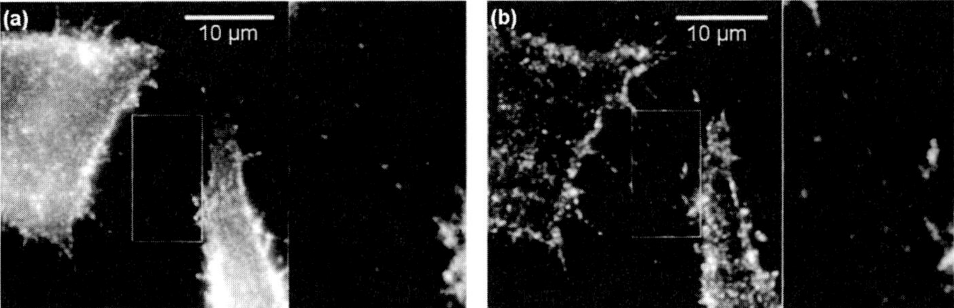

Figure 6.5: Room-temperature TIRF-M and epi-fluorescence images of CHO cells incubated in hypo-osmotic PBS. Imaging of filopodia of CHO cells stably transfected with the SNAP-D_{2L} receptor, incubated in hypo-osmotic PBS (108 mOsm) for 2 h and then labeled with Alexa546-BG in epi-fluorescence **(a)** and TIRF-mode **(b)**. The inserts show higher magnification images of the areas in the white boxes. (Images courtesy of Alina Tabor.)

thorough cleaning procedure. We verified that the HF treatment did not influence the diffusion and dimerization of SNAP-D_{2L} receptors. In this experiment (c.f. Fig. 6.3 for a schematic drawing of the optical and cryogenic setup), the microscope objective was located outside of the vacuum chamber, so that objective-based TIR illumination was not accessible. Instead, we used a wide-field illumination, which was accompanied with a considerably higher background level caused by out-of-focus fluorescence compared to samples fabricated by spin-coating extremely clean solutions of purified molecules. Because of the larger absorption cross-sections of chromophores at cryogenic temperatures this effect is particularly dramatic in our experiment. To address this issue, we reduced the amount of cellular material in the region of interest by inducing filopodia formation through exposure of the cells to hypo-osmotic stress conditions (see Fig. 6.5). The direct comparison between wide-field illumination and TIRF-mode for the same CHO cells at room temperature showed that the level of background fluorescence in the filopodia regions is similar. We, thus, concentrated our analysis on these regions during the cryogenic measurements.

6.3.1 Bayesian localization analysis

The recorded data in this experiment at cryogenic temperatures still showed fairly high background fluorescence and a high number of blinking and overlapping spots, not allowing us to apply standard single molecule localization algorithms. There have been several attempts to deal with higher labeling densities and the resulting overlapping fluorophores, e.g. super-resolution optical fluctuation imaging (SOFI) [263, 264] and the Bayesian localization analysis (Bayesian bleaching and blinking, 3B analysis) [265]. The latter technique models each emitting fluorophore as a Gaussian spot with the properties: position (x, y), radius, and intensity. The algorithm then performs a global optimization using Bayesian inference to find optimal values for these properties as well as to determine the number of emitters. This way, blinking fluorophores that reappear at a later, not necessarily adjacent, frame will have an improved localization precision. The principle of the 3B analysis is briefly described in the following paragraph.

Let $\mathcal{F}$ be the hypothesis that a fluorophore exists at a certain position, and $\mathcal{N}$ that no fluorophore exists given the data, $\mathcal{D}$. The goal of the optimization is to determine the relative conditional probabilities [265]

$$\frac{P(\mathcal{F}|\mathcal{D})}{P(\mathcal{N}|\mathcal{D})} = \frac{P(\mathcal{D}|\mathcal{F})P(\mathcal{F})}{P(\mathcal{D}|\mathcal{N})P(\mathcal{N})} \quad , \tag{6.1}$$

where we used Bayes' theorem. The probabilities $P(\mathcal{F})$ and $P(\mathcal{N})$ are constants, and the calculation of $P(\mathcal{D}|\mathcal{N})$ is trivial since it is the probability of observing all data given the noise model. Computation of $P(\mathcal{D}|\mathcal{F})$ is considerably more complex, in particular since the number of fluorophores M is unknown and models with different fluorophore number $\mathcal{F}_M$ must be compared. Computation of $P(\mathcal{D}|\mathcal{F}_M)$ employes a series of numerical and statistical techniques such as Markov Chain Monte Carlo (MCMC) sampling and Maximum-a-posteriori estimation, and is computationally costly [265,266]. Additionally, MCMC is known to have difficulties in estimating a termination point for the optimization [266], which is why one usually runs a large number of iterations until there are no more significant changes in the result [265].

We performed the 3B analysis using the ImageJ PlugIn provided by Rosten and coworkers [267]. Relatively large data sets (50×50 px up to 70×70 px, 1,000 frames) were processed using default settings and with measured values for the PSF FWHM. The analysis ran for > 200 iterations to ensure convergence (runtime > 7 days on an Intel i7 3.4 GHz workstation).

6.3.2 Cryogenic localization microscopy of dopamine receptors

Figure 6.6a shows the time-averaged recording of two CHO cells at cryogenic temperatures. A region with filopodia is located between the two cells (white square). Applying the 3B analysis to this region yielded the super-resolution reconstruction image shown in Fig. 6.6b. The localization precision was determined to be better than 10 nm by analyzing cross sections.

From the homology model of a dopamine D_2 receptor homodimer based on the β_1 adrenergic receptor, attached to two SNAP-tags, we expected the distance of the centers of mass to be about 5 nm (see Fig. 6.4). To extract distance information from our experimental data, we computed pairwise distances from the positions of the fluorophores as determined by the 3B analysis. Figure 6.6c shows a histogram of the pairwise distances (Euclidian metric) for the SNAP-D_{2L} dopamine receptor (blue) calculated from the identified positions. We corrected the histogram for an offset that stems from random distances

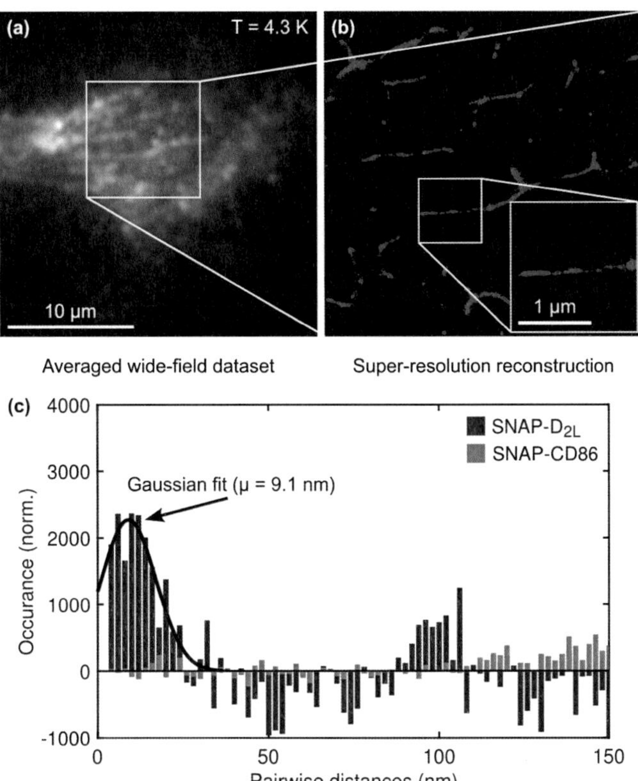

Figure 6.6: Cryogenic localization microscopy of dopamine receptor dimers. **(a)** Averaged wide-field dataset of a recording of a CHO cell stably expressing labeled SNAP-D_{2L}. **(b)** Super-resolution reconstruction after 3B analysis of the area indicated by the white square in (a). **(c)** Histogram of the pairwise distances calculated from the emitter positions of the SNAP-D_{2L} (blue) and SNAP-CD86 monomers (gray). The Gaussian fit determines the separation of the SNAP-D_{2L} protomers to be $\mu = 9.1 \pm 11.3\,\text{nm}$.

due to either background fluorescence that was mistakenly identified as a fluorophore or receptors in a monomeric state. The histogram was also corrected by subtracting a simulated histogram which was computed in the following way: The same number of emitters was randomly placed on a three-dimensional cylinder of 150 nm diameter [268] and with the accumulated filopodia length. Pairwise distances were then histogramed after computing a two-dimensional projection.

As a control experiment, we looked at CHO cells stably expressing SNAP-CD86 monomers (gray histogram in Fig. 6.6c). There is a clear peak in the histogram for SNAP-D_{2L} (blue) at 9.1 ± 11.3 nm which is not present in the control experiment with the monomeric membrane proteins (gray). The value of 9 nm for the distance between the two receptor protomers is in good agreement with the protomer separation as expected from the structural alignment.

6.4 Summary

The size of dopamine receptors is approximately 5 nm, about 40 times smaller than the diffraction limit of light. In order to distinguish between random co-localization of two closely-lying receptors and the physical interaction of two protomers in a receptor dimer, we applied cryogenic localization microscopy. After inducing filopodia formation through exposure of the cells to hypo-osmotic stress conditions, SNAP-D_{2L} receptors in intact CHO cells were frozen using cryo-immobilization, and their behavior was compared to monomeric CD86 membrane proteins.

Cryogenic localization microscopy allowed us to display SNAP-D_{2L} receptor dimers at high resolution and to measure the distance between their centers of mass. The determined distance of about 9 nm indicated a physical interaction of the protomers. The experimentally determined distance is in good agreement with a homology model of the dopamine D_2 receptor homodimer.

In the following chapter, we verify the feasibility cryogenic distance measurement with Ångström accuracy by measuring the separations of organic dye pairs attached to a small double-stranded DNA.

7 Co-localization of two fluorophores at nanometer separation

The spatial resolution of localization microscopy is mainly limited by the number of detected photons. In chapter 5 we have shown that cryogenic measurements improve the photostability of fluorophores, giving access to Ångström precision in localization of single molecules.

In this chapter, we extend this method to the co-localization of two fluorophores attached to well-defined positions at the backbone of a double-stranded DNA. By measuring the separations of the fluorophore pairs prepared at different design positions, we verify the feasibility of cryogenic distance measurement with Ångström accuracy. We discuss the important challenges of our method as well as its potential for further improvement and various applications.

The content of this chapter has been published as:
S. Weisenburger, B. Jing, D. Hänni, L. Reymond, B. Schuler, A. Renn, and V. Sandoghdar, *Cryogenic Colocalization Microscopy for Nanometer-Distance Measurements* ChemPhysChem **15**, 763–770 (2014).

Passages of the present text might be nearly identical to the text in the published manuscript.

7.1 Introduction

An important step beyond single-molecule localization is co-localization of several fluorophores and demonstration of Ångström resolution. There is a plethora of applications where Ångström resolution is desirable. In particular, it would be highly interesting to perform quantitative distance and orientation measurements in a single biomolecule or in nanometer-sized molecular complexes.

To co-localize two or several molecules within a diffraction-limited spot, it is necessary to detect single emitters sequentially [135]. Soon after it became possible to detect single molecules [39, 101], co-localization of single molecules was demonstrated in early 1990s by selective excitation of narrow transitions within the inhomogeneous distribution of molecular resonance frequencies at liquid helium temperature [136, 242]. In 2002, cryogenic high-resolution single-molecule spectroscopy was used to demonstrate nearly molecular resolution in all three spatial dimensions by creating position-dependent frequency shifts through the application of electric field gradients [8]. Extension of this spectral selectivity to room temperature has also been explored using multicolor nanocrystal quantum dots [138] or fluorescent molecules [13] but it is hampered by the broad overlapping emission spectra and chromatic aberration in the optics. Stochastic photoactivation strategies employed in room-temperature experiments [141, 269] can also be applied in cryogenic measurements [270] to improve their resolution. An interesting possibility to avoid restriction to photoactivable labels is, however, to exploit step-wise bleaching or blinking [9, 10, 271, 272]. Here, discrete levels in the fluorescence trace of a molecule are identified and the corresponding frames are analyzed to localize each PSF.

In chapter 5 we have demonstrated that the significantly reduced rate of photochemistry at cryogenic temperatures allows single-molecule localization at Ångström precision. In this chapter, we extend this exquisite precision to determine the separation of two neighboring fluorophores on the backbone of a double-stranded DNA as a model system.

7.2 Materials and methods

7.2.1 Sample preparation

A DNA strand with the sequence GCGAGTTCCACCTACCCTGCCTAAGCCTG-TATC(C6dT)GTCA was labeled at position 34, where C6dT represents the modified thymidine deoxynucleosides with a flexible linker containing six methylene groups and a terminal amine group. This strand was annealed to different labeled oligonucleotides with each construct resulting in a different sequence separation. The sequence of the complementary second strand was CGCTCAAGGTGGATGGGACGGATTCGGACATA-GACAGT with the nucleotide at either position 4, 14, 20, or 24 replaced by C6dT. The first strand additionally contained a biotinylated poly-A tail for surface immobilization for room-temperature experiments that were not discussed here. Modified oligonucleotides (Microsynth AG) were purified with ion-exchange chromatography and labeled with Alexa Fluor 532 succinimidyl ester (Invitrogen). The two strands were then hybridized to obtain a double-stranded DNA. Since double-helical DNA has a persistence length of about 50 nm, we expect our short DNA constructs (less than 15 nm long) to behave like rigid rods [217].

The samples for microscopy were prepared by spin coating the labeled DNA constructs for 10 s at 1000 rpm followed by 30 s at 3000 rpm. A buffer solution with 130 µl Tris-EDTA buffer (Fluka, BioUltra (10 mM Tris-HCl; 1 mM EDTA; pH 7.4)) was prepared with 5 µl of 1 M $MgCl_2$ (Sigma (anhydrous, $\geq 98\,\%$)). Next, 10 µl of a diluted solution of about 0.5 µM of the DNA constructs was added. 3 µl of this stock solution was then spin coated on fused silica cover slips (thickness 170 μm, Esco Products) that were cut to about $7 \times 7\,\mathrm{mm}^2$ square pieces before they were thoroughly cleaned by alternating oxygen plasma and rinsing with deionized water as well as non-halogenated solvents (acetone, ethanol, methanol, and 2-propanol, in that order). The samples were placed in the cryostat chamber immediately after preparation.

7.2.2 Cryogenic and optical setup

Technical details of the experimental setup and the localization fitting procedure can be found in chapter 2 and 4. In short, the experiments were performed on a homebuilt epi-fluorescence microscope, NA = 0.75, with the sample mounted in a liquid helium flow cryostat. We minimized mechanical vibrations and drifts by a rigid and centro-symmetric setup design, and waiting for the setup to be settled after the initial cooling down and every mechanical movement. Furthermore, we accounted for a remaining long-term drift on the order of 100 nm per hour by tracking fiduciary markers.

7.3 Results and analysis

In order to co-localize two or several molecules within a diffraction-limited spot, it is necessary to detect the individual emitters sequentially. Because the absorption spectra of common dye molecules do not reduce beyond the inhomogeneous broadening of the spectrum, spectral selection via high-resolution spectroscopy is not possible in these systems [39,101]. An interesting alternative is to analyze the PSF corresponding to discrete intensity levels of stepwise bleaching or blinking of single molecules [9,10,271,272]. Here, we follow the latter strategy.

First, we extract a fluorescence intensity trace for a single PSF from an image stack recorded at a frame rate of about 5 Hz. This frame rate turned out to be a good compromise between a reasonable time resolution to capture fast blinking events and a SNR of $\approx$ 50 per frame, which is high enough for reliable localization. Figures 7.1a and 7.1c plot two different examples of blinking traces, showing discrete levels of intensity, corresponding to both or one of the molecules fluorescing. We then identify the intensity levels (solid lines in Fig. 7.1a) using a total variation-based denoising [273]. Total variation denoising is an efficient method for edge-preserving reduction of noise from step-like signals. The used algorithm minimizes the biased discrete total variation functional. We also plot histograms of the fluorescence intensities as displayed in Figs. 7.1b and 7.1d for the traces of Figs. 7.1a and 7.1c, respectively.

Once we have identified the blinking intervals, we sum the frames within each one to improve the SNR. We then determine the position x_1 of one of the molecules from the PSF of a low-intensity interval. An example of the diffraction-limited spot corresponding to the interval marked by an arrow in Fig. 7.1a is displayed in Fig. 7.3a. Figure 7.3b shows

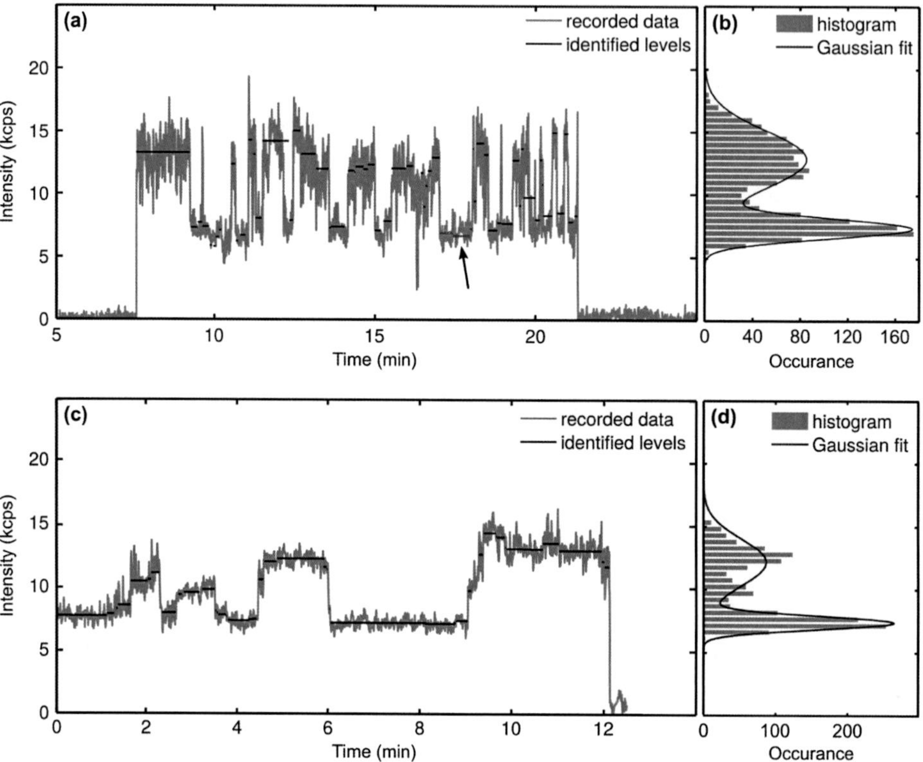

Figure 7.1: Blinking behavior at cryogenic temperatures. (a) Example fluorescence trace of a single DNA construct labeled with two Alexa molecules. The arrow marks the interval (red) for which the PSF is shown in Fig. 7.3a – b. **(b)** Histogram of the fluorescence intensities (gray) and a double Gaussian fit (blue). **(c,d)** Same as (a,b) but for a different DNA construct.

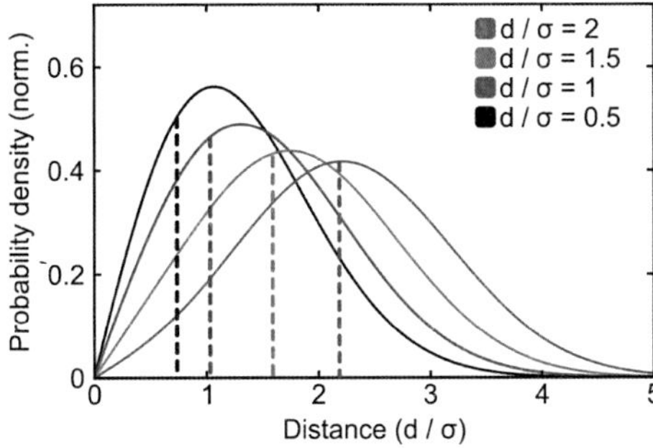

Figure 7.2: Illustration of the asymmetric distance distribution. Examples for the Euclidian norm of bivariate normal distributions for different ratios d/σ between 0.5 and 2. Dashed lines mark the positions of the expectation value.

a line cut from this PSF as well as a fit that yields a localization precision of 7 Ångströms (corresponding to a SNR of $\approx 3{,}000$). Next, we find the PSF center of mass $\boldsymbol{x}_c$ of the fluorophore pair from one of the neighboring high-intensity intervals and compute the distance between the two molecules by accounting for the fluorescence intensities of the two molecules in a weighted fashion. The position $\boldsymbol{x}_2$ of the second molecule is then

$$\boldsymbol{x}_2 = \frac{N_c \boldsymbol{x}_c - N_1 \boldsymbol{x}_1}{N_c - N_1} \quad , \tag{7.1}$$

where N_1 and N_c denote the number of photons of the first molecule and the center of mass spot, respectively.

The two localized positions $\boldsymbol{x}_1$ and $\boldsymbol{x}_2$ are each normally distributed in space with localization precisions σ_1 and σ_2. The distance vector $\boldsymbol{d} = \boldsymbol{x}_1 - \boldsymbol{x}_2$ is normally distributed about $\boldsymbol{\mu} = \boldsymbol{x}_1^{\text{true}} - \boldsymbol{x}_2^{\text{true}}$ with variance $\sigma^2 = \sigma_1^2 + \sigma_2^2$. Since the distance, however, is a non-negative number, the Euclidian norm $d = \|\boldsymbol{x}_1 - \boldsymbol{x}_2\|$ cannot have a Gaussian distribution. In order to estimate $\mu = \|\boldsymbol{\mu}\| = \|\boldsymbol{x}_1^{\text{true}} - \boldsymbol{x}_2^{\text{true}}\|$ from measurements of $d = \|\boldsymbol{d}\|$, we can write the Gaussian probability distribution $p(\boldsymbol{d})$ in polar coordinates [274],

$$\begin{aligned} p(\boldsymbol{d}) &= p(r,\phi) && (7.2)\\ &= \frac{1}{2\pi\sigma^2} \exp\left\{ -\frac{(\boldsymbol{d}-\boldsymbol{\mu})^2}{2\sigma^2} \right\} && (7.3)\\ &= \frac{1}{2\pi\sigma^2} \exp\left\{ -\frac{(\mu^2 + d^2 - 2r\mu\cos\phi)^2}{2\sigma^2} \right\} \quad . && (7.4) \end{aligned}$$

Integration over ϕ yields

$$\begin{aligned} p(d) &= d \int_0^{2\pi} \mathrm{d}\phi\, p(r,\phi) && (7.5)\\ &= \left(\frac{d}{\sigma^2}\right) \exp\left\{ \frac{\mu^2 + d^2}{2\sigma^2} \right\} I_0(r\mu/\sigma^2) \quad , && (7.6) \end{aligned}$$

where I_0 denotes the modified Bessel function of the first kind and order zero.

Figure 7.2 plots some examples of distance distributions which are the Euclidian norm of a bivariate normal distribution. This distribution is also called a Rice distribution [275]. Importantly, the expectation value of this asymmetric distribution does not coincide with the peak position so that a Gaussian fit would introduce a systematic error towards a larger distance.

As shown by several examples in Fig. 7.3c – e for different DNA constructs, we plot the extracted distance values from various interval pairs of each blinking trace in a histogram and perform a maximum likelihood estimation with the Rice distribution to obtain mean and accuracy values. The results are displayed by the fits in Fig. 7.3c – e.

All stated localization precision values were computed from the covariance matrix of the fitting procedure by error propagation of the variance of the residuals. Individual frames where the localization precision was worse than the design distance, e.g. for very short blinking intervals, were discarded. We note that the variations in the number of distance readings in the examples of the histogram stem from different observed blinking behavior, e.g. shown in Figs. 7.1a and 7.1c.

Figure 7.4a sketches the example of a DNA construct with a nominal distance of 10 nm, for which we find a distance of 11.8 ± 1.2 nm between the two fluorophores. Figure 7.4b shows the results for DNA constructs of four different design fluorophore separations of 10 nm (30 base pairs, bp), 6.7 nm (20 bp), 5.0 nm (14 bp), and 3.3 nm (10 bp). The open circles in Fig. 7.4b denote individual measurements, and the error bars show the standard error of the mean. The data clearly confirm the ability of cryogenic co-localization to read the distances of single molecules in the range well below ten nanometers. The variation in the error bars is caused by the stochastic nature of photoblinking and the thus variations in the SNR of the individual measurements. We mention in passing that one has to keep in mind that the size of an Alexa Fluor 532 molecule itself (molecular weight of 720 Da) is approximately 1.3 nm and it is attached to the nucleoside via a linker of six methylene groups. This uncertainty in position due to label and linker size is statistically averaged out due to symmetry when looking at multiple DNA molecules.

7.4 Discussion

We now discuss a few intriguing features of the physics and open questions involved in our experiments. Curiously, the fraction of DNA molecules that showed more than one discrete intensity level in their blinking trace was much lower than expected. For instance, in the case of the DNA constructs with 10 nm separation we recorded about 12,000 traces from wide-field image stacks. Only less than 10 % of these showed two distinct levels of fluorescence intensity. We emphasize that in room-temperature experiments we verified that this low yield was not caused by suboptimal labeling. After excluding the traces, where either two separate fluorophores accidentally lay close by (i.e. the

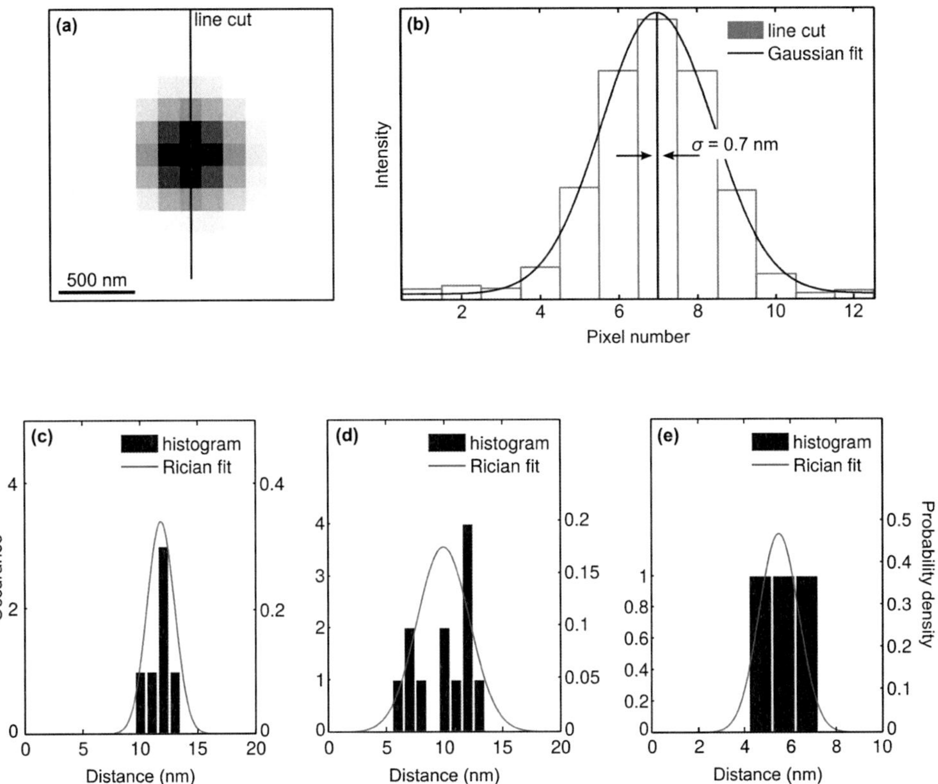

Figure 7.3: Co-localization analysis. (a) Diffraction-limited spot on the camera chip from the interval marked in Fig. 7.1a. **(b)** Line cut (gray) and Gaussian fit (blue). The black line illustrates the position of the center determined with a precision of 0.7 nm. **(c – e)** Three examples of histograms of the distances determined from individual steps between adjacent intervals of the blinking traces (blue). The green curves show fits according to a Rice distribution to determine the mean distance and the accuracy of the distance measurements.

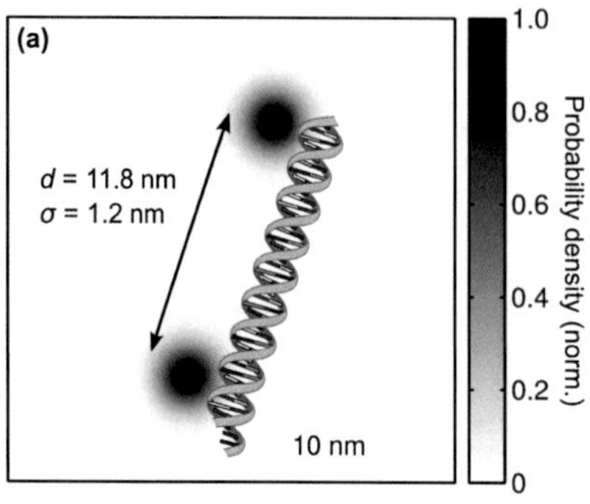

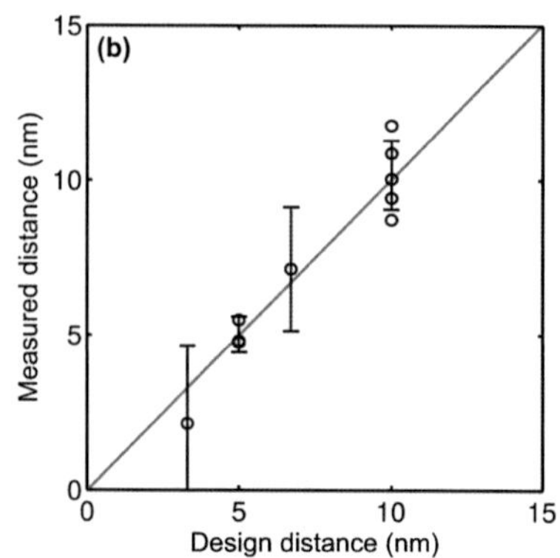

Figure 7.4: Cryogenic co-localization measurements. (a) Visualization of the mean positions and localization precision for a DNA construct with two Alexa Fluor 532 labels placed at a separation of 10 nm. **(b)** Results of DNA constructs with four different fluorophore design distances of 10 nm (30 base pairs, bp), 6.7 nm (20 bp), 5.0 nm (14 bp), and 3.3 nm (10 bp). Each open circle denotes a measurement of one fluorophore pair. Error bars show the standard error of the mean.

extracted distance was well above the design value) or the SNR was too low, only 0.1 % of the original traces could be used for co-localization. For DNA constructs with smaller fluorophore separation this fraction was even lower.

A second noteworthy finding is that we never observed total off-states (i.e. both fluorophores off) in the blinking traces. Assuming a scenario where the off-state is caused by a photo-induced charge separation and subsequent charge trapping in the vicinity of the fluorescent molecule [276], a low number of possible trapping sites or a filled trap could explain this observation. Furthermore, the distribution of intensities for the higher level was always broader than that of the lower level by 2 – 3 times. While a factor of $\sqrt{2}$ can be attributed to a larger shot noise, we do not have a robust explanation for this discrepancy.

Another interesting observation is that blinking off and on states last for tens to hundreds of seconds, whereas the same samples display much faster blinking at room temperature. These long residence times facilitate the high localization precisions reported in this work. Finally, we did not encounter two-step bleaching events. In other words, no recorded trace showed a transition from two active fluorophores to a state, where only one fluorophore was alive for a long time.

The photophysics and photochemistry of nearby molecules can be complex. Various phenomena such as dipole-dipole coupling and fluorescence energy transfer [161,227,277] as well as nontrivial distribution of charges and chemically-active elements such as oxygen might be at play. A proper understanding of these processes requires extensive spectroscopic studies which are beyond the scope of this work. Nevertheless, the data in Fig. 7.4b indicate that reliable information can be extracted even from a small number of runs although more measurements would also clearly reduce the measurement uncertainty.

In the experiments reported here, the DNA strands carrying the fluorophores lay at the interface so that the effective separation of the centers of mass of the dipoles and the interface was of the order of one nanometer. Furthermore, we could only use a microscope objective with a low numerical aperture of NA = 0.75. As a result (see also chapter 4), dipole emitters with out-of-plane orientation were less efficiently excited, and large-angle components of the emission were not captured. It turns out, therefore, that the localization accuracy of our measurements is not compromised in this arrangement.

7.5 Summary

In chapter 5 we reported Ångström precision in single-molecule localization at cryogenic temperatures. Here, we have extended those results to co-localization of two identical fluorescent molecules placed at nanometer distances. The aim of this study has been to demonstrate the feasibility of cryogenic co-localization for measuring separations of molecules in the range well below ten nanometers. This method is particularly promising for quantitative distance measurements in systems such as membrane receptor stoichiometry [257, 278, 279], protein folding [280, 281], protein conformational dynamics using temperature cycling [282] or orientation determination of biomolecules [45]. Moreover, the technique can be used to obtain information on the position of light emission in multichromophore systems such as light-harvesting complexes or j-aggregates [283, 284].

In the next chapter we extend our methodology to the co-localization of multiple fluorophores attached to a small protein.

8 Light microscopy of protein structure at Ångström resolution

Insight into the atomic and molecular structure of proteins and other biomolecular assemblies is highly desirable in many areas of life sciences, and several physical techniques such as X-ray crystallography, electron microscopy (EM) and magnetic resonance spectroscopy have been employed over decades to obtain such information. However, each approach has its limitations and, therefore, additional methods are in demand.

In this chapter, we present a novel optical microscopy technique, termed Cryogenic Optical Localization in three Dimensions (COLD), which reaches Ångström resolution in deciphering the positions of several fluorescent sites within a single small protein. We determine the conformational state of the cytosolic PAS domain from the histidine kinase CitA of *Geobacillus thermodenitrificans*. Furthermore, we resolve the four sites where biotin binds to streptavidin. In both cases, the comparison to the crystal structure of the proteins shows an excellent agreement.

The content of this chapter is part of the following manuscript:
S. Weisenburger, D. Boening, B. Schomburg, K. Giller, S. Becker, C. Griesinger, and V. Sandoghdar,
Light Microscopy of Protein Structure at Angstrom Resolution
Nature Methods, under review.

Passages of the present text might be nearly identical to the text in the submitted manuscript.

8.1 Introduction

Inspired by the contributions of cryogenic electron microscopy [216] and the success of room-temperature super-resolution microscopy, we have pursued low-temperature investigations, where orders of magnitude more photons can be collected before photobleaching sets in (see chapter 5). In previous chapters, we demonstrated single-molecule localization at Ångström precision (chapter 5) and cryogenic distance measurements by resolving two fluorophores on a double-stranded DNA (see chapter 7). In this chapter,

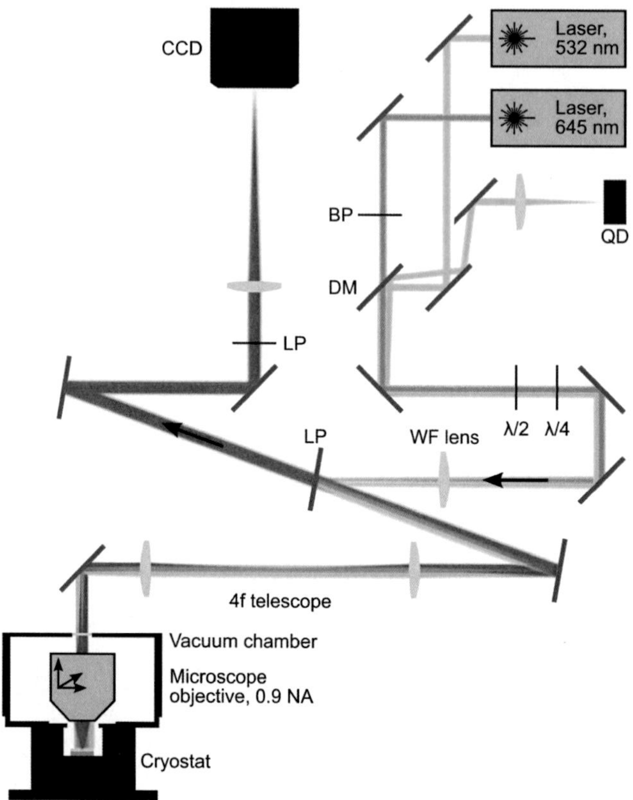

Figure 8.1: Schematic of the cryogenic and optical setup. Schematic of the beam path and optical components. BP - band pass filter; DM - dichroic mirror; QD - quad diode; $\lambda/2$ - half wave plate; $\lambda/4$ - quarter wave plate; WF lens - wide-field lens; LP - long pass filter; CCD - CCD camera.

we show that this method can resolve the three-dimensional (3D) positions of several fluorophores attached to a protein.

8.2 Materials and methods

8.2.1 Optical and cryogenic setups

Technical details of the experimental setup can be found in chapter 2. A schematic of the experimental setup for the measurements presented in this chapter is shown in Fig. 8.1a. We used a diode laser at wavelength λ=645 nm for excitation and passed it through a 645-10 bandpass filter (Semrock). This beam was combined with another laser beam at

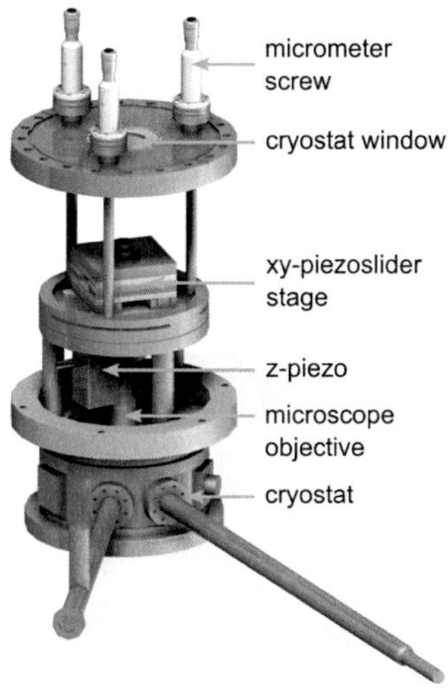

Figure 8.2: Cryostat setup used for COLD microscopy. A commercial flow cryostat (lower part of the drawing) is extended by a vacuum chamber (upper part; the walls are not shown for clarity) houses a microscope objective with a numerical aperture of 0.9. The sample is placed on a cold finger under the objective. See Methods for details.

λ=532 nm by a dichroic mirror. The sample was illuminated by circularly polarized light generated by two wave-plates. A wide-field lens (f = 400 mm) focused the laser beam via a 4f telescope (f = 350 mm) into the backfocal plane of the microscope objective.

The microscope objective (Mitutoyo, 0.9 NA LWD) was mounted inside a homebuilt extension to the cryostat vacuum chamber (see Fig. 8.2 for a schematic view). The microscope objective was mounted on a piezo-electric objective positioner for fine axial motion (MIPOS, Piezosysteme Jena), and it could be moved laterally by a custom-built 2D piezo-slider stage (Smaract). For coarse vertical adjustment, the slider system was placed on a platform that could be adjusted by three fine-threaded micrometer screws via vacuum feed-throughs working against a plate spring. The sample (see below) was glued on a home-built gold-plated OFHC copper cold finger using Apiezon N.

For sample cooling we operated a liquid helium flow cryostat (Janis ST-500) at liquid helium temperature. The laser power was adjusted to 0.2 mW before the cryostat window. The fluorescence was detected in epi-mode. A 650 nm longpass filter at low angle of incidence (about 5 degrees) was used as a beam splitter. A second 650 nm longpass filter was used as a detection filter before a f = 300 mm lens focused the light onto a water-cooled CCD camera. The physical pixel size was 16 µm, and the magnification of the imaging system was 182 times, translating to a pixel size of 88 nm. Wide-field image stacks were recorded at a frame rate of 0.5 Hz in frame transfer mode with EM gain switched off and using the 1 MHz analogue-digital converter. Typical recordings had 3,600 frames where the acquisition was interrupted every 15 min for automated refocusing using the 532 nm laser reflection from the substrate interface and a quadrant photodiode for a feedback loop.

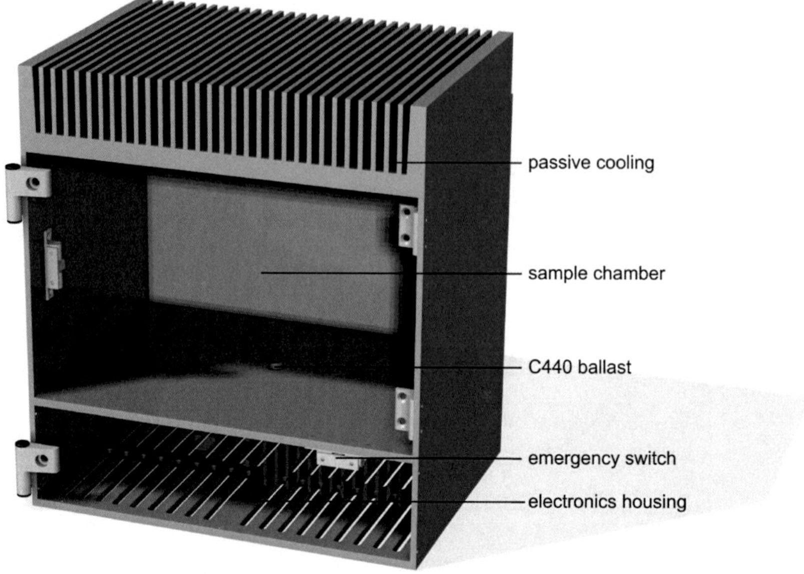

Figure 8.3: Schematic of the home-built bleaching chamber. The ballast has a socket to mount the high pressure mercury lamp.

8.2.2 Cover slip cleaning and bleaching chamber

UV-grade fused silica cover slips (ESCO) were cut to 7.0 × 7.0 × 0.2 mm pieces and then thoroughly cleaned using the following protocol: The cover slips were rinsed with non-halogenated solvents (acetone and ethanol) and stored in Piranha solution (H_2SO_4 and H_2O_2, ratio 1:1). Before usage, they were ultrasonicated in Helmanex (4 % at 60 °C) for one hour, rinsed with ethanol and plasma cleaned (O_2 plasma, 10 min). Then they were ultrasonicated in HCl and H_2O_2 solution (ratio 3:1) at 60 °C for one hour. Next, they were rinsed with 2-propanol followed by another 10 min in the plasma cleaner (O_2 plasma).

In order to reduce the background luminescence in the recorded images, a bleaching chamber was engineered and constructed, where sample substrates can be stored and exposed to intense UV-VIS light. Figure 8.3 shows a schematic view of the bleaching chamber. The chamber housing is made from aluminum and the inner chamber has a volume of about 25 l for sample storage. It also includes electronics for a safety switch that is connected to the (double) door and the C440 ballast (Heraeus) which has a socket to mount a high pressure mercury lamp (HPL 125W, Heraeus). The lamp features several spectral peaks between 250 nm and 600 nm with about 2 Wm^2nm^2 spectral irradiance.

Figure 8.4 demonstrates the effect of the bleaching chamber on background lumines-

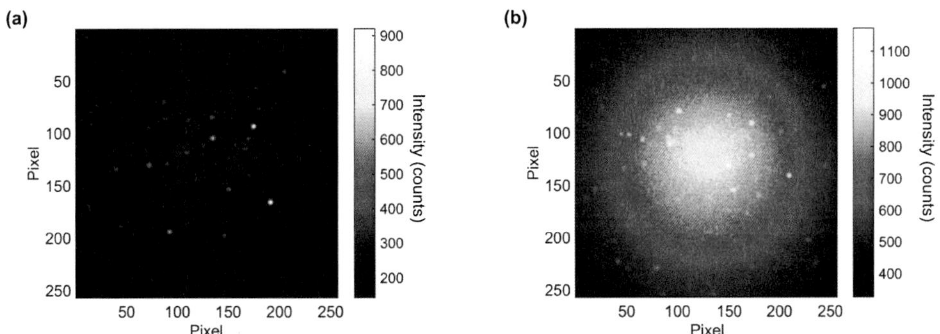

Figure 8.4: Effect of the bleaching chamber on background luminescence. Camera frame recorded at cryogenic temperature from a single molecule sample where the cover slip has been placed in the bleaching chamber for about 6 h prior to sample preparation **(a)** and comparison without pre-bleaching **(b)**.

cence from the used cover slips. Panel (a) shows a camera frame, integration time 2 s, of a sample containing single Atto 647N (ATTO-Tec) molecules spin coated on a cover slip that was placed for about 6 hours in the bleaching chamber prior to sample preparation. As a comparison, panel (b) shows an example frame with the same conditions of a similar sample where the cover slip was cleaned in the same fashion but not placed in the bleaching chamber. We generally observe for our cover slips a reduction of the background luminescence of a factor of three to four at cryogenic conditions after placing the cover slip in the bleaching chamber.

8.2.3 Protein labeling and sample preparation

For the conjugation of streptavidin (Sigma) with biotin-Atto647N (Sigma) both proteins were incubated in Tris-EDTA buffer (pH 7.4) at concentrations of 0.8 M for biotin-Atto647N and 16 mM for streptavidin for two hours at room temperature. After the reaction, the solution of bound and unbound biotin was purified using a centrifugal concentrator (Vivaspin 20, Satorius) with a cutoff at 30 kDa. The degree of labeling was determined by UV-VIS absorption spectroscopy using a Nanodrop-2000 (Thermo Scientific). We used Tris-EDTA buffer for blanking and took an extinction coefficient of $\epsilon_{280} = 167,000\,\mathrm{M^{-1}cm^{-1}}$ at 280 nm. The extinction coefficient for Atto647N is given in the data sheet as $\epsilon_{\lambda} = 150,000\,\mathrm{M^{-1}cm^{-1}}$ and the correction coefficient as $C_{280} = 0.05$ (AttoTec GmbH). The degree of labeling (DOL) was determined to be about DOL = 4.

CitA PASc (residues 200 – 309) from *Geobacillus thermodenitrificans* was modified by site-directed mutagenesis to carry a C-terminal cysteine residue (N308C) which could be used for dye conjugation. Uniformly ^{15}N-labeled CitA PASc N308C was expressed in *Escherichia coli* BL21(DE3) cells in M9 minimal medium with ^{15}N-ammonium chloride as

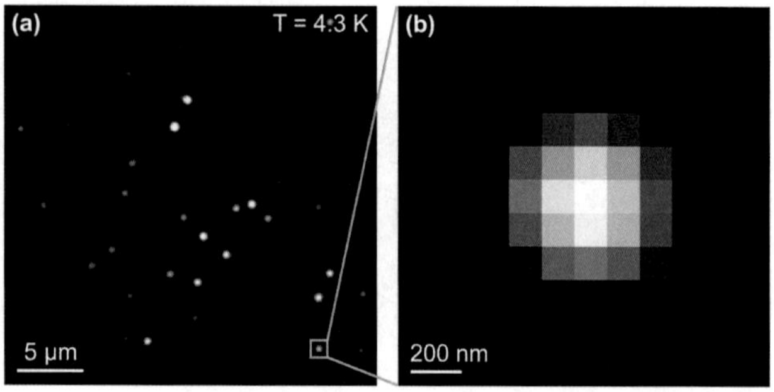

Figure 8.5: Example of a wide-field fluorescence image. (a) Camera frame recorded at liquid helium temperature with an integration time of 2 s. **(b)** Zoom on one diffraction-limited spot.

nitrogen-source. After induction with IPTG, cells were incubated in a shaking culture for 5 hours at 30°C before harvesting. Cell pellets were re-suspended in lysis buffer (20 mM Tris · HCl pH 7.9, 300 mM NaCl, 10 mM imidazole, 0.5 mM phenylmethylsulfonylfluoride (PMSF)) and ruptured by sonication. PASc N308C protein was collected via immobilized metal affinity chromatography on Ni-NTA resin (Qiagen). The N-terminal His-tag was cleaved by incubating the protein with TEV protease. Cleaved PASc N308C was reloaded onto Ni-NTA resin and the flow-through of the resin dialyzed against size exclusion chromatography (SEC) buffer (20 mM Na-phosphate pH 6.5, 150 mM NaCl). The final SEC purification was performed with a Superdex 75 16/60 column (GE Life Sciences). The purified PASc N308C sample was dialyzed against phosphate buffered saline (PBS), pH 7.4 and subsequently concentrated to 200 μM. For the conjugation of the PASc domain with Atto647N-maleimide (AttoTec GmbH) we followed the standard protocol (AttoTec GmbH) and removed unbound dye by gel filtration on a SD75 10/30 column. The protein was then exchanged to 10 mM Tris buffer at pH 7.4 and subsequently spin-concentrated. We estimated the concentration by UV-VIS spectroscopy to be about 15 μM using an extinction coefficient of $\epsilon_{280} = 38,400\,\mathrm{M}^{-1}\mathrm{cm}^{-1}$ at 280 nm. The DOL for the PASc domain was determined to be about 1.3.

To prepare the sample, we spin coat the proteins in an approximately 100 nm thin polymer layer on a small cover slip. A solution for spin-coating was prepared by mixing 90 μl of Tris-EDTA buffer, 20 μl polyvinyl alcohol (PVA) (10 %w, steril filtered, degassed), 10 μl Trolox (20 mM) in DMSO and 10 μl Streptavidin-Biotin Atto647N (aliquots from −20 °C resuspended in 1 ml Tris-EDTA buffer, then diluted 1:100 in Tris-EDTA) or PASc stock solution (aliquots from −80 °C, then diluted 1:10,000 in Tris-EDTA), respectively. 5 μl of this solution was spin-coated on the cover slip (10 s at 1000 rpm, 60 s at 3000 rpm). The concentration of proteins on the cover slip is adjusted such that their average distance is much larger than the system PSF. Figure 8.5a shows an example of a diffraction-limited fluorescence frame that was recorded with an exposure time of 2 s, and Fig. 8.5b displays a zoom into the PSF of one bright spot from panel (b). After spin coating, the sample was immediately transferred to the cryostat and cooled down.

8.2.4 Single-molecule localization analysis

Technical details of the localization fitting procedure can be found in chapter 4. To determine the positions of the individual neighboring emitters attached to a given protein, they must be localized sequentially. Here, we take advantage of the naturally occurring stochastic photoblinking, which takes place on the time scale of seconds and minutes at cryogenic temperatures. Although N identical labels would give rise to N possible brightness levels, variations in orientation, local environment and quantum efficiency of the fluorophores lead to 2^N combinations of the on/off-state signal levels. We fitted such intensity time traces to a model, which comprises of the possible linear combinations of the expected fluorescence signal I,

$$I(t) = \sum_{k=1}^{N} b_k(t) I_k \quad , \tag{8.1}$$

for all time points t, to estimate the individual intensity levels I_k. Here, $b_k(t) \in \{0,1\}$ denotes the on/off state of signal $k = 1 \ldots N$. The error of the fit was calculated by a least squares estimation with the square root of the recorded intensity as a weight. We only used the coordinates corresponding to the lowest intensity levels, i.e. of individual fluorophores for our localization data.

We also checked our localization analysis for systematic errors (see chapter 4). Briefly, due to the fact that we could only use a microscope objective with a low numerical aperture of NA = 0.9, dipole emitters with out-of-plane orientation were less efficiently excited, and large-angle components of the emission were not captured. Thus, it turns out that the localization accuracy of our measurements is not compromised in this arrangement.

8.2.5 Three-dimensional reconstruction

We can localize each fluorophore that contributes to the recorded images, but the random orientation of the protein in the sample leads to a continuous distribution of the projections onto the detection plane. To account for the large number of projection possibilities, we follow the protocol of single-particle reconstruction used in electron microscopy [170].

For 3D reconstruction we prepared localization images by placing 2D Gaussian functions at the localized positions with the respective localization precisions as width parameters (100 × 100 Å, 1.0 Å mesh grid). In case of streptavidin, we only used data sets where all localization precisions were better than 13 Å and the maximal distance of two localizations was below 80 Å to avoid cropping artifacts in the reconstruction. For experiments with PASc, all localization precisions were better than 6.5 Å. The single particle reconstruction was performed using the subspaceEM algorithm, a fast maximum-a-posteriori expectation-maximization (E-M) algorithm, from reference [285] with default

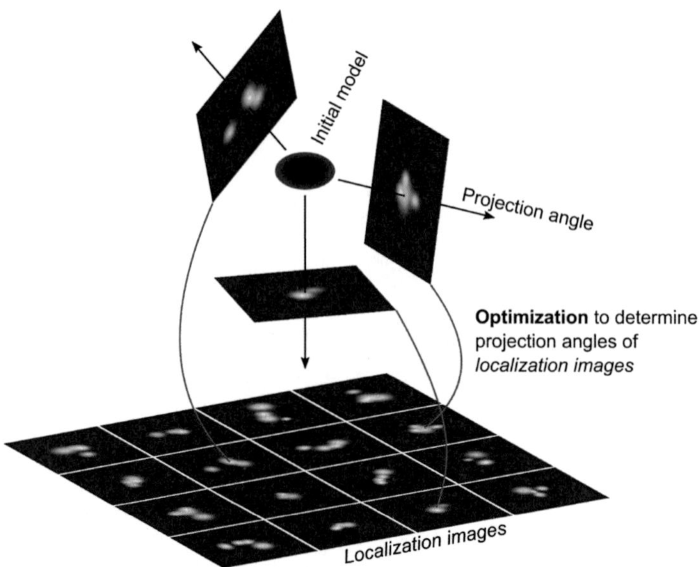

Figure 8.6: Schematic of the EM reconstruction algorithm. The EM algorithm tries to determine the optimal orientation angles of the projection images. In the second step, a new model is computed from the projection images. The algorithm iterates these two steps until it converges.

settings except for the number of iterations and the number of runs which were set to 25 and 100, respectively. To avoid any bias in the reconstruction, the initial model is an ellipsoid with $a, b = 10$ Å and $c = 20$ Å placed in the center of the volume.

The E-M algorithm is an established algorithm to infer a maximum likelihood estimate when the likelihood function contains incomplete data. These latent variables are in our case the unknown projection operators and image transformation operators, i.e. it is unknown how the projection images are rotated and translated, and along which direction they were projected. Figure 8.6 shows a schematic that explains how the E-M algorithm works. It iterates between the expectation step and the maximization step. First, latent probabilities for every possible alignment of the recorded images to the projection of the reconstructed volume were calculated. Then, in the maximization step, these probabilities were used to generate weights for the alignments in order to compute the reconstruction.

As a check experiment we falsely applied the four-fluorophore model to the data of single Atto647N molecules and performed the reconstruction. As expected, the 3D reconstruction yielded one spherical object in the center of the volume (c.f. Fig. 8.7).

We also checked if the algorithm could reconstruct a volume where we generated artificial data to have a ground truth. For this purpose we placed four 3D Gaussian functions in the volume and created 100 projection images at random orientations. The

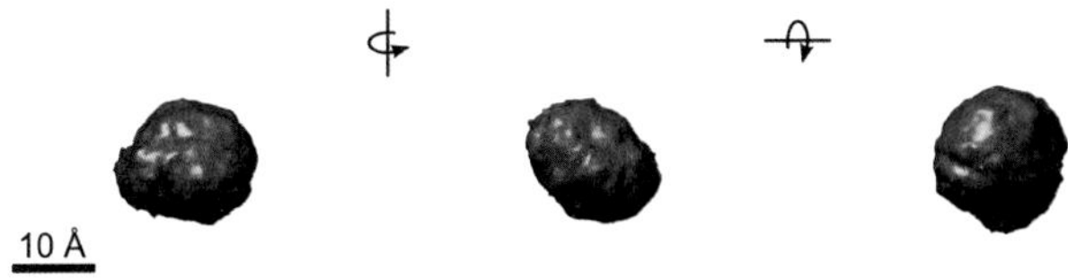

Figure 8.7: Reconstruction of Atto647N. 3D reconstruction from localization images that were generated from the single fluorophore (Atto647N) data under the false assumption that each spot comprises four fluorophores. The reconstruction yields a single point volume.

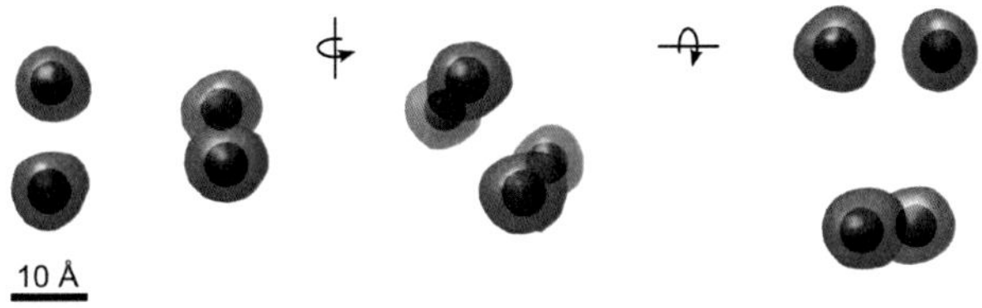

Figure 8.8: Reconstruction of artificial data. 3D reconstruction (red) from 100 projection images that were generated from four 3D Gaussian functions placed in the volume (black). For better visualization the black spheres are plotted at 50 % of their original diameter.

reconstruction worked nearly perfectly (see Fig. 8.8).

A quantitative value for the resolution of our 3D reconstruction was estimated by calculating the Fourier shell correlation (FSC) of two half data set reconstructions [286]. The 3D FSC determines the normalized cross-correlation coefficient of the two independent reconstructions over corresponding shells in Fourier space [287],

$$\mathrm{FSC}(r) = \frac{\sum_{r_i \in r} F_1(r_i) \cdot F_2(r_i)^*}{\sqrt{\sum_{r_i \in r} |F_1(r_i)|^2 \cdot \sum_{r_i \in r} |F_2(r_i)|^2}} , \tag{8.2}$$

where F_1 denotes the complex structure factor for volume 1 and F_2 the one for volume 2, and r_i are the individual Fourier space voxels in shell r. The resolution is then determined by finding the point of intersection of the FSC curve with the curve of a resolution criterion. We used the half-bit criterion which states a resolution where the reconstructed volume contains sufficient information to interpret it [287]. The FSC was computed using Free FSC (IMAGIC).

8.2.6 Modeling of the crystal structure

In order to compare our results with the protein structure, we attached the four biotin ligands by modeling to the crystal structure of streptavidin (PDB ID: 1STP). We constructed

	Streptavidin	PASc domain dimer
Identified proteins	6,239	2,133
After filtering	114	188
Yield	1.8 %	8.8 %

Table 8.1: Overview of the recorded data and the experimental yield. We have performed about 60 similar experiments over a period of seven months. The used data include 8,372 single protein measurements. Data were filtered by all localization precisions better than 13 Å (streptavidin) and 6.5 Å (PASc domain dimer).

a 3D structure of Atto647N based on its structure as described in reference [288] using MolEditor to generate Smiles and Avogadro to build and optimize the 3D geometry. Using UCSF Chimera [289], we coupled Atto647N fluorophores to the biotin ligands. For the PASc domain, the Atto647N dyes were attached to the N308C positions (mutated to Cystein) at the C-termini of the crystal structure (PDB ID: 5FQ1) by modeling. We used the point-symmetry of the volume maps of the 3D reconstruction and fitted them to the crystal structure using UCSF Chimera, which was also used for visualization.

8.3 Results and discussion

8.3.1 Cytosolic PAS domain of GtCitA

As a first step, we studied the sensor histidine kinase CitA from the thermophilic bacterium *Geobacillus thermodenitrificans* (Gt), which regulates the transport and anaerobic metabolism of citrate in response to its extracellular concentration [290]. GtCitA consists of an extracytoplasmic citrate-binding Per-Arnt-Sim (PAS) domain flanked by two transmembrane helices, a cytosolic PAS domain (PASc), which is a 12.2 kDa dimeric protein domain, and a conserved kinase core. We attached two Atto647N dyes to the two C-termini of the cytosolic PAS domain.

By fitting the resulting intensity time trace to a model, which comprises of 2^N possible linear combinations, the individual intensity levels were estimated. Figure 8.9a displays an example of the observed brightness levels for the PASc domain. The red trace in Fig. 8.9a presents an example of the outcome, where indeed $2^2 = 4$ intensity levels were clearly identified. The shot-noise-limited fit error plotted in Fig. 8.9b emphasizes the large confidence with which each intensity is assigned. To co-localize the two fluorophores, we determined the lateral coordinates of the PSFs corresponding to the observed two lowest intensity levels by weighted least-square fitting with a 2D Gaussian function. The gray histogram in Fig. 8.9c presents all individual localization precisions. For an efficient analysis, we only used data sets (blue histogram) where both fluorophores of the protein had a localization precision better than 6.5 Å (see Table 8.1 for experimental yields).

Figure 8.9d presents a few examples recorded projections for separate proteins that

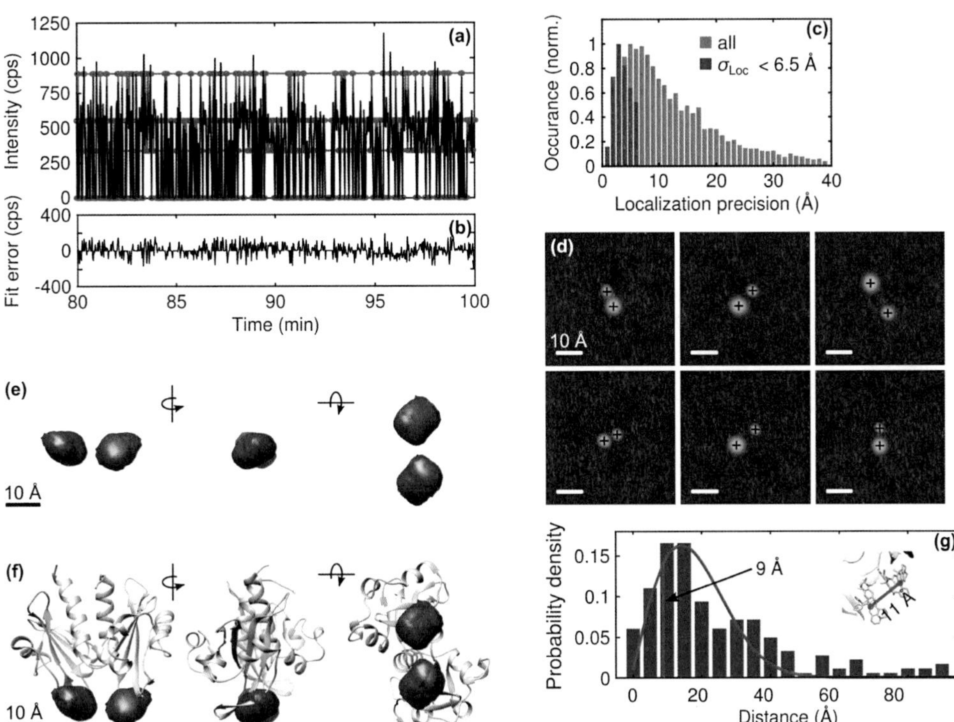

Figure 8.9: 3D reconstruction of the GtCitA PASc domain dimer. **(a)** Exemplary intensity trace (black line) of two fluorophores with linear combination fit (red dots). **(b)** Plot of the weighted fit error as the difference between the linear combination fit and the intensity trace. **(c)** Histogram of the localization precisions of all recorded data (gray) and data used for further analysis (blue). **(d)** Examples of recorded projections for separate proteins. Black crosses indicate the measured positions while the size of the yellow circles denote the precision in each case. **(e)** Reconstructed volumes of the fluorophores plotted at an isovalue of 0.68. Views from three orthogonal directions are shown. **(f)** Overlay of the reconstructed volumes (red) with the crystal structure. **(g)** Histogram of the measured distances and the expected distribution (red curve) as determined by a model fit. The red vertical line shows the resulting expectation value. See Methods for details. The inset illustrates the expected relative locations of the two fluorophores as determined from the crystal structure.

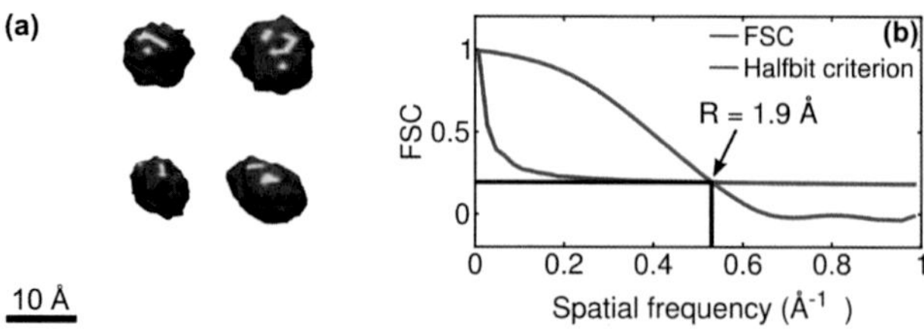

Figure 8.10: Fourier shell correlation of two half data sets for the PASc domain dimer. (a) 3D reconstructions of two half data sets. **(b)** The halfbit criterion yields a resolution of 1.9 Å.

have different orientations in the sample. In the next step, we perform the described single-particle reconstruction to arrive at a self-consistent 3D solution without any *a priori* knowledge of the investigated structure. Figure 8.9e presents the isosurfaces of the reconstructed probability densities at an isovalue of 0.68. The overlay of these results with the crystal structure of the PASc dimer (PDB ID: 5FQ1) reveals a very good agreement (see Fig. 8.9f). To quantify the spatial resolution of the procedure, we computed the Fourier shell correlation [286] of two reconstructions from half data sets (c.f. Fig. 8.10a,b). Comparison with the half-bit criterion [287] yields a resolution of 1.9 Å (see Fig. 8.10c).

It is evident that our measurement resolves the two labels within a single protein. In order to extract more quantitative information from our data, in Fig. 8.9g we plot a histogram of the measured distances together with a fit based on a Monte Carlo model and simulated annealing [291], taking into account our localization precision (the red curve). Simulated annealing is a probabilistic method for finding the global minimum of a complicated cost function with local minima. By allowing for occasional moves to a parameter set increasing the cost function value instead of decreasing it, the solver is able to escape local minima and explore a larger space of solutions. The possible cost function increase is determined by the current *temperature* parameter. In the course of the optimization, the temperature is lowered according to an annealing schedule. The method was named simulated annealing because of the similarity of this procedure with annealing of a metal. The asymmetric distribution stems from the convolution of a projection function and the Euclidian norm of a bivariate normal Rice distribution [275] (see chapter 7). As a consequence of this skewed distribution, the expectation value does not coincide with the peak position. The obtained distance of 9 ± 3 Å between the two fluorophore sites is in excellent agreement with the expected distance of 10.9 ± 2.1 Å for the dyes from the protein crystal structure (see Fig. 8.9g, inset).

The rare occurrences of larger distances in Fig. 8.9g amount to about 10 %. A closer scrutiny of the outliers reveals abnormal intensity time traces, which prevent a robust identification of the lowest levels for localization. This could be in turn caused by rare photophysical events such as short bursts of intense fluorescence emission caused by sudden changes in the nano-environment of the dye molecule, or imperfections in the

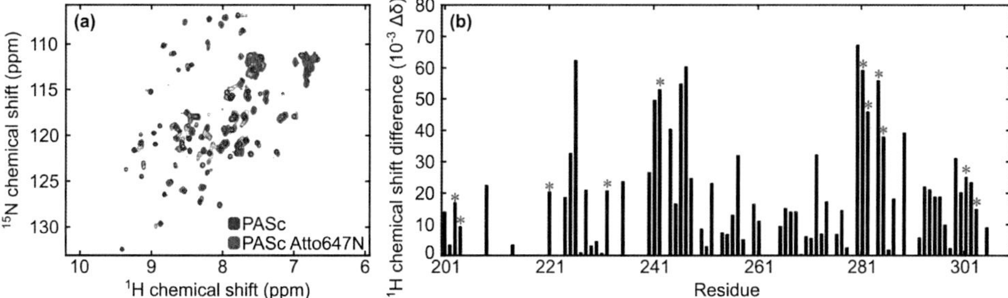

Figure 8.11: NMR measurements on labeled PASc. (a) Overlay of ^{1}H, ^{15}N-HSQC-spectra of 1 mM GtCitA PASc (blue) with 25 μM Atto647N-labeled PASc N308C (red). Peaks for unfolded PASc after dye labeling were not observable. Peaks corresponding to the β-scaffold are partially broadened out due to anisotropic effects from the dye molecules. **(b)** Proton chemical shift difference of identified residues in Atto647N-labeled PASc compared to unlabeled PASc reference shifts. Only minor chemical shift changes are visible. An observed peak doubling for some residues (red asterisks) is likely caused by the degree of labeling, DOL = 1.3, resulting in different molecular species in the NMR sample to be detected. (Images courtesy of Benjamin Schomburg.)

labeling procedure. Instead of sorting out these data points, however, we have chosen to include them to demonstrate the robustness of our procedure.

Furthermore, by using NMR measurements, we have verified that the presence of labels does not significantly alter the protein structure. Our collaborators recorded ^{1}H, ^{15}N-HSQC-spectra of GtCitA PASc on a Bruker 800 MHz spectrometer equipped with a 5 mm CPTCI cryoprobe at 298 K (see Fig. 8.11a). The chemical shifts induced in Fig. 8.11b agree with expectations for the extended conjugated system.

The 5 Å long 6C linker of the dye label can contribute to the experimental uncertainty although the specific local protein environment restricts the flexibility of the linker orientation to a large extent.

8.3.2 Streptavidin Atto647N-biotin conjugates

Next, we increased the complexity of COLD by considering a protein that is labeled with four fluorophores. We chose the streptavidin homo-tetramer with a molecular weight of about 52.8 kDa and conjugated it with four Atto647N-biotin ligands. Figures 8.12a shows an example of an intensity time trace for such a streptavidin conjugate. Again, the small uncertainty in fitting 16 intensity levels (see Fig. 8.12b) displays the robustness of the procedure. Figure 8.12c plots a histogram of all localization precisions and the ones for the data sets that were used for further analysis (see Table 8.1 for experimental yields), and Fig. 8.12d presents some examples of the recorded projections.

Figure 8.12e shows the outcome of our 3D reconstructed volumes, and Fig. 8.12f overlays these with the crystal structure (PDB ID: 1STP) of the streptavidin conjugate.

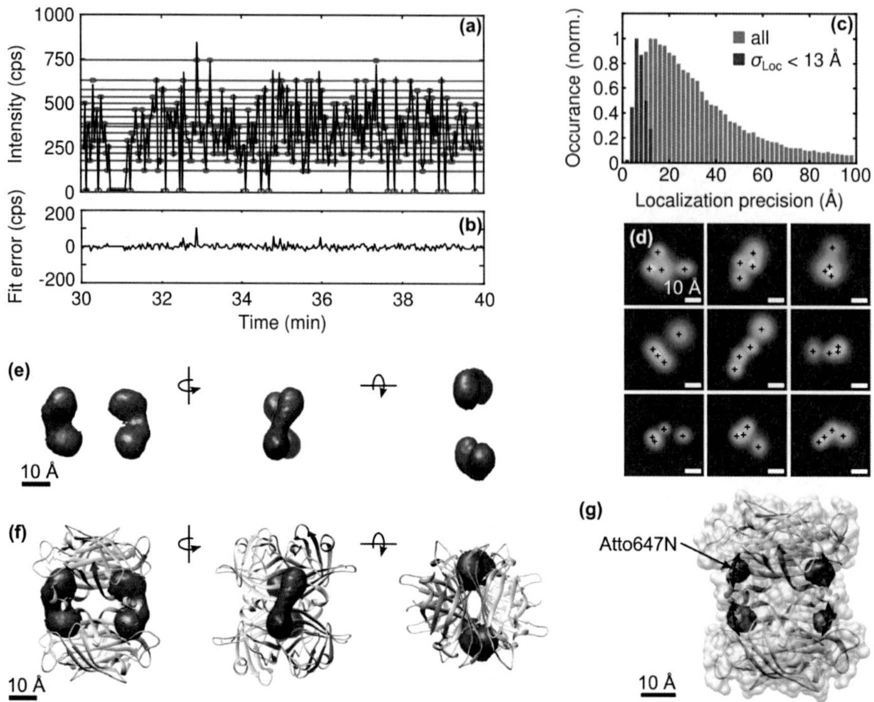

Figure 8.12: 3D reconstruction of the streptavidin-biotin binding sites. **(a)** Exemplary intensity trace (black line) of four fluorophores with linear combination fit (red dots). **(b)** Plot of the weighted fit error as the difference between the linear combination fit and the intensity trace. **(c)** Histogram of the localization precisions of all recorded data (gray) and data used for further analysis (blue). **(d)** Examples of recorded projections for separate proteins. Measured positions are indicated by black crosses. **(e)** Reconstructed volumes of the fluorophores plotted at an isovalue of 0.68. Views from three orthogonal directions are shown. **(f)** Overlay of the reconstructed volumes (red) with the crystal structure. **(g)** Same as in (f) but together with the electron surface of the streptavidin and the fluorophore volumes set at an isovalue of 0.80. The arrows indicate the possibility of linker rotation.

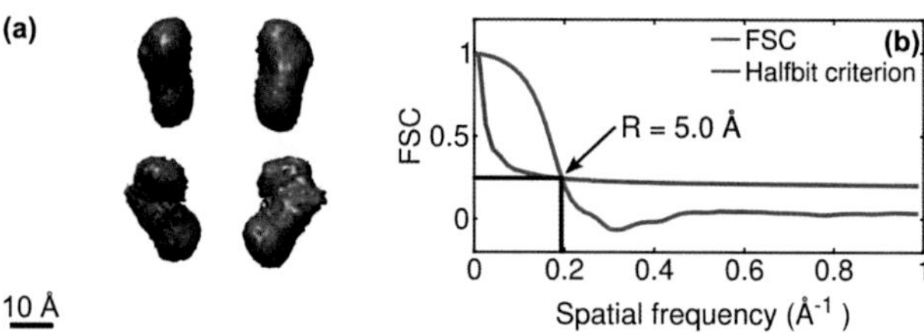

Figure 8.13: Fourier shell correlation of two half data sets for the streptavidin complex. **(a)** 3D reconstructions of two half data sets. **(b)** The halfbit criterion yields a resolution of 5.0 Å.

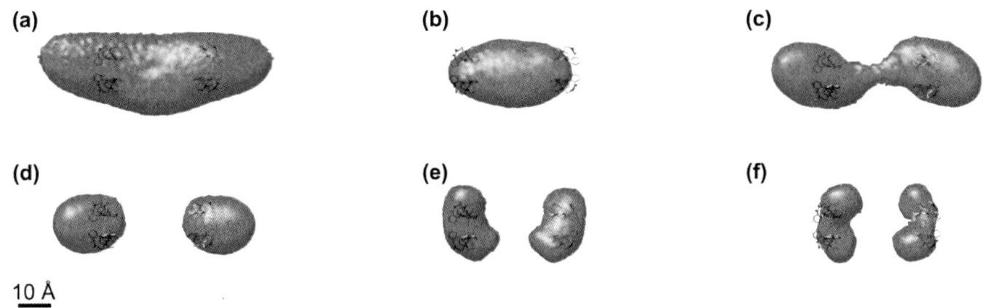

Figure 8.14: Increasing quality of the 3D reconstruction with better input data. **(a)** 12 Å $< \sigma_{\text{Loc}} <$ 30 Å; **(b)** 10 Å $< \sigma_{\text{Loc}} <$ 25 Å; **(c)** 8 Å $< \sigma_{\text{Loc}} <$ 22 Å; **(d)** 6 Å $< \sigma_{\text{Loc}} <$ 20 Å; **(e)** $\sigma_{\text{Loc}} <$ 17 Å. **(f)** $\sigma_{\text{Loc}} <$ 13 Å. Isosurfaces are plotted at an isovalue of 0.68.

The Fourier shell correlation indicates a resolution of 5.0 Å in this case (see Fig. 8.13). We attribute the slight concave curvature of the reconstructed volumes towards the inside of the protein to the rotational freedom of the flexible linker that allows it to pivot about its attachment point within a certain solid angle (see also Fig. 8.12g). Interestingly, as displayed by Fig. 8.12e, f, our measurements also reveal a systematic small (< 10Å) offset of the biotin pockets with respect to each other.

Furthermore, we checked how the quality of the 3D reconstruction increases when we increase the quality of the input data. In order to do so, we increased the requirement on the localization precision for the input data while keeping the number of localization images constant. We started with using the data where the localization precision σ_{Loc} was 12 Å $< \sigma_{\text{Loc}} <$ 30 Å (see Fig. 8.14a) and gradually increased the requirement up to $\sigma_{\text{Loc}} <$ 17 Å (Fig. 8.14e). As we can see, with better input data the quality of the reconstruction increases dramatically. Only at 8 Å $< \sigma_{\text{Loc}} <$ 22 Å the two volumes start to separate (Fig. 8.14c).

8.3.3 Orientation of individual proteins

The high resolution and the tomographic nature of COLD make it possible to identify the orientation of the protein in the laboratory frame and with respect to its environment. To examine this, we compared our individual localization measurements to the projections of a model based on the crystal structure decorated with four fluorophores modeled as 3D Gaussian spheres with a sigma value of 5 Å. Here, we accounted for the experimental uncertainties as the width of the Gaussian volumes by including the measured uncertainty of 5 Å (see Fig. 8.9c) and allowing for another 5 Å as the maximum error caused by the flexible linker. These model projections were then fitted to our measured localization images by a least-squares fit using the Euler angles of the orientation of the model volume as fit parameters. Figure 8.15 shows some examples. The nearly perfect agreement between the experimental data (Fig. 8.15a–d) and model projections (Fig. 8.15e–h) is

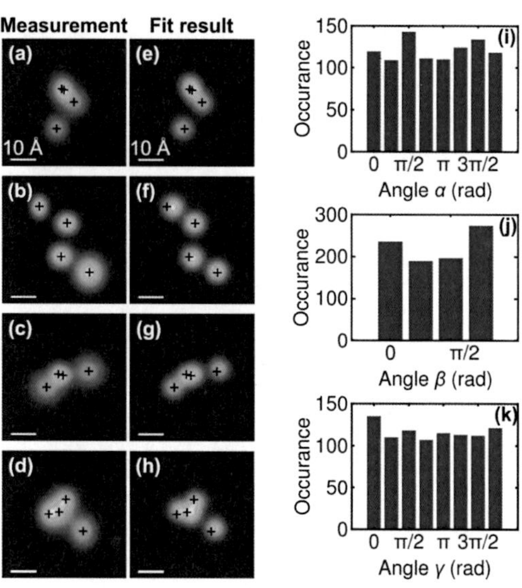

Figure 8.15: Determination of single protein orientation. Four examples of localization projections (**a–d**) and the results of least-squares fitting to the projections of a model (**e–h**). The uncertainty in the measured position is visualized by the width of 2D Gaussian functions. The black crosses mark the experimental positions in both cases. **i–k,** Histograms of the extracted three Euler angles α, β and γ.

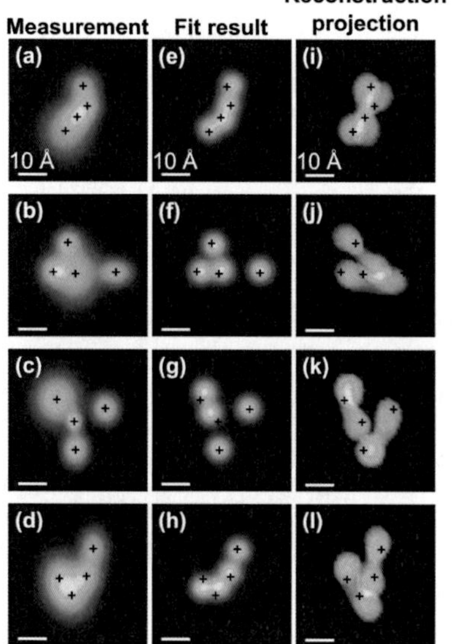

Figure 8.16: Individual protein orientation and fit to the 3D reconstruction. Additional data sets with measurements **(a–d)**, least-squares fit to the projection of a model **(e–h)** and to projections of the 3D reconstruction volume **(i–l)**. The uncertainty in the measured position is visualized by the width of 2D Gaussian functions. The black crosses mark the experimental positions in both cases.

remarkable. The histograms for the obtained three Euler angles (see Fig. 8.15i–k) indicate that in our case the proteins were, indeed, oriented randomly within the polymer film. Figure 8.16 shows additional datasets of single molecule orientation fits and also fits to the projections of the 3D reconstruction volume. Again, the experimental data show excellent agreement with the projections of the reconstructed volumes.

8.4 Summary

In this chapter, we presented our results on resolving the positions of multiple fluorophores attached to proteins using COLD microscopy. By applying algorithms borrowed from cryogenic electron microscopy, we could reconstruct a three-dimensional density map for the positions of the fluorescent labels with a resolution of several Ångström, yielding excellent agreement with the expected crystal structure. Our technique pushes optical resolution by nearly two orders of magnitude beyond the state-of-the-art room-temperature super-resolution microscopy. It allows us to gain structural information that might not be accessible via existing analytical methods such as X-ray crystallography or magnetic resonance spectroscopy.

By reaching Ångström optical resolution, fluorescence microscopy confronts its fundamental limit, where the dimensions of the label and its linker become the limiting factor. A great future promise of COLD is in investigating proteins and molecular assemblies. Often such structures are part of a larger arrangement of identical molecules which makes it difficult to label with different fluorophores as it is required in Förster Resonance Energy Transfer (FRET). This is the case in the two examples that were shown in this chapter.

When compared with other methods, it can be stated that COLD is more sensitive than electron paramagnetic resonance (EPR) spectroscopy or pulsed electron double resonance (PELDOR) [292], which also rely on identical paramagnetic labels in oligomeric structures. However, techniques based on paramagnetic resonance require orders of magnitude larger sample amounts and have a precision that is limited by the flexibility of the paramagnetic label similar to our method. Moreover, soluble as well as membrane proteins even in complex with further recognition partners can be studied with our technique.

9 Conclusion and outlook

9.1 Concluding remarks

The endeavor to devise new imaging methods and push the spatial and temporal resolution is a fundamental challenge for physicists and of great practical importance in science and technology. Ernst Abbe's formulation of the diffraction limit at the end of the nineteenth century put a harsh spell on optical microscopy, which lasted for about one hundred years. The inception of scanning near-field microscopy broke this spell, and once the dogma of a fundamental limitation of the optical resolution was eliminated, scientists reconsidered many ideas and explored fascinating techniques that have revolutionized optical imaging. Several seminal methods were invented in the past two decades which improved the diffraction-limited resolution in optical imaging by one order of magnitude to about 20 nm. These super-resolution fluorescence microscopy techniques, which were awarded the Nobel Prize in Chemistry in 2014, allowed for resolving sub-cellular structures and organelles, and have enabled discoveries in neuroscience, molecular biology and other life sciences.

In the course of this dissertation, we developed a new microscopy technique that exploits cryogenic measurements to push the optical resolution further by another two orders of magnitude to the Ångström level. The near molecular resolution is made possible by the substantial improvement of the molecular photostability at liquid helium temperature. We described the constructed experimental setups, the background on the photophysics of the used fluorescence markers as well as the analytic methods. We introduced our cryogenic super-resolved fluorescence microscopy technique and demonstrated single-molecule localization with Ångström localization precision. We also presented an application of our methodology to whole cells, where we could determine the distance between the protomers of receptor homodimers. Next, we verified the feasibility of co-localization and cryogenic distance measurements by resolving two fluorophores at nanometer separation attached to a double-stranded DNA.

Finally, we presented our results on resolving the positions of multiple fluorophores attached to a small protein using cryogenic localization microscopy. By applying algorithms borrowed from cryogenic electron microscopy, we could reconstruct a three-dimensional density map for the locations of the fluorescent labels with an unprecedented resolution of a few Ångström, yielding excellent agreement with the expected crystal structure. Our technique establishes a new record of the three-dimensional resolution in optical microscopy by pushing it nearly one hundred times beyond the state-of-the-art room-temperature super-resolution microscopy. It allows us to gain structural information that might not be accessible via existing analytical methods such as X-ray crystallography or magnetic resonance spectroscopy. This constitutes a decisive breakthrough in both optical imaging and structural analysis of matter.

9.2 Future prospects

Improving the localization precision All results that we presented would benefit from even higher precision and accuracy than it was demonstrated here. One possible way to increase the localization precision is to use a microscope objective with higher numerical aperture. In addition to a smaller point-spread function size, this will result in an increase in the collection efficiency and a higher number of detected photons. For example, the use of a solid immersion lens made of gallium phosphide with refractive index of about $n = 3.4$ would increase the number of detected photons by 4.5 times [209] and enhance the localization precision by about $3.4\sqrt{4.5} \approx 7$ folds, taking optical microscopy beyond the Ångström resolution. As discussed above, in this case one would have to consider the systematic errors that arise from the orientation of the dipole emitters close to an interface. Algorithms accounting for these effects have already begun to surface [231,293]. A practical disadvantage of using a single hemispherical lens is the limited field of view. This challenge can be addressed by using micro-lens arrays which offer multiple fields of view [294].

An increased detection efficiency also leads to another effect; If nearly all emitted photons are detected, the photon noise becomes sub-Poissonian. Using a planar metallo-dielectric antenna, more than 99 % collection efficiency has been experimentally demonstrated [223]. Such a collection efficiency holds the promise for a noise level of $\sqrt{1 - 0.99} = 0.1$ times shot noise if one saturates the molecule, allowing for an even further improvement of the localization precision [295].

Dipole-dipole coupling Co-localization of several emitters relies on the distinguishability of each individual one. If neighboring emitters couple through near-field dipole-dipole interaction [8, 277], one ceases to obtain separate point-spread functions from each individual molecule but rather expects a collective point-spread function. The success of our measurements indicates that this effect does not play a role in our system. The conditions for the dipole-dipole coupling to perturb co-localization depend on several factors such as the relative orientation of the dipole moments, absorption and emission spectral linewidths, or fluorescence quantum yields. Although we have successfully demonstrated that resolving several close-lying fluorophores by optical microscopy is feasible, there are still many open questions concerning the photophysics of these fluorophores, in particular at cryogenic temperatures [8]. We will address these questions in studies to come.

Size of the fluorescent label and the linker Cryogenic measurements take fluorescence microscopy to its ultimate limit with a promise for sub-Ångström resolution: The dimensions of the label and the flexible linker are becoming the limiting factor. Our results provide incentives to engineer more rigid and shorter bonds for fluorophore attachment [296,297]. An alternative strategy for reducing the effects of the linker is to restrict the accessible volume that the fluorescent label can explore by a careful choice of the attachment point, which has been explored in the context of Förster Resonance Energy Transfer [298]. A possibility to avoid the problem of label and linker altogether is to exploit the autofluorescence of molecules. At room temperature, the intrinsic autofluorescence signal from amino acids such as tryptophan is very weak [299], but the enhanced photostability at low temperatures can enable label-free structure studies of proteins using cryogenic localization microscopy.

Correlative and complementary character Cryogenic localization microscopy can be combined with and complement the established analytical techniques commonly used in structural biology, and it can provide valuable insight that may not be accessible by those. It also ideally lends itself to correlative microscopy with other methods such as cryogenic electron microscopy or magnetic resonance spectroscopy. The unmatched specificity of fluorescent labeling can be used to resolve ambiguities or to identify an analyte molecule in the presence of high concentrations of other molecules. Our method also provides three-dimensional information about the position of the fluorescent labels allowing for direct comparison to 3D structures determined by other structural biology techniques. They usually rely on measuring large numbers of identical molecules. In contrast to that, cryogenic localization microscopy is a single-particle method and delivers spatial information about the structure of molecules at very low concentrations. In addition to soluble molecules, we also expect to be able to study membrane proteins in their native complex arrangements with other recognition partners and in different conformational states.

Classification algorithms The use of classification algorithms [170] will allow for measuring samples which contain different protein species or proteins in different conformational states simultaneously. This may play an important role in experiments which rely on high throughput to accumulate sufficient statistics. Class averaging has already been combined with standard super-resolution microscopy to obtain high resolution images of structures which otherwise would have insufficient signal-to-noise ratio [300,301].

Studying dynamic processes A strategy to access dynamic and transient information at cryogenic temperatures is to freeze and measure a process at different time points. It should also be possible to explore dynamic protein folding by combining our method with local *in situ* heat cycling [302]. By using a focused near-infrared laser one can raise and lower the temperature of small sample volumes by almost two hundred Kelvin within microseconds.

Vitrification and cryo-transfer An intriguing application of our methodology is to perform stochastic optical reconstruction microscopy (STORM) of sub-cellular structures at cryogenic temperature. Chemical fixation can result in structural changes of the sample [303], especially on the ultra-structural level in cells which is the most interesting for super-resolution imaging [304, 305]. Naturally, the best possible preservation of structures can be achieved when performing super-resolution imaging with live cells. There have been efforts to accelerate the acquisition of super-resolution images, but live-cell super-resolution microscopy still remains challenging [306–308]. An alternative to live-cell imaging is fixation of the sample by vitrification. This entails cryo-immobilizing the structure in a glassy amorphous ice by flash freezing techniques. This way, structures can be preserved in a near-native state as it is already applied in other fields such as cryogenic electron microscopy or X-ray microscopy [309, 310].

We have demonstrated the feasibility of applying cryogenic localization microscopy to whole cells. Our experiments using Bayesian localization analysis can be extended by the implementation of a cryo-transfer for vitrified samples. We also performed some preliminary cryogenic STORM experiments using the recently introduced silicon rhodamine tubulin label (SiR-Tubulin) [311]. SiR-Tubulin is membrane permeable and therefore allows live-cell labeling of tubulin structures without the need of expressing fluorescent proteins. We achieved an average localization precision of about four nanometers; The excellent performance and the compatibility of SiR-Tubulin with live-cell labeling and vitrification makes this system a promising candidate for further investigations. It also has a considerably smaller probe size compared to immunofluorescence staining, which is particularly important for achieving high labeling density of the structure of interest, which is key to high-resolution super-resolution imaging.

A concrete topic in forefront research is to shed light on the spatial and temporal organization of genetic material within the nucleus of the cell [312]. Optical microscopy has revealed the morphology and dynamics of the nucleus but it has not been able to resolve the three-dimensional organization of DNA at the sub-nucleosome level [313,314]. Classical ligation methods such as chromosome conformation capture have provided a first glimpse into genome architecture [315], but they only provide information at select loci or on a genome-wide scale. Since chromosome organization is highly dynamic, correlation of high-spatial resolution structure with gene activity will play an essential role. Readily available techniques such as fluorescence *in situ* hybridization [313] and gene editing using CRISPR/Cas9 [316] can be used for specific labeling. Although hampered by the limited resolution, first room-temperature super-resolution studies to investigate the 3D structure of chromatin have begun to emerge [317]. Cryogenic localization microscopy holds the promise for Ångström spatial resolution for directly visualizing 3D organization of chromosomes and the nucleus in general.

It is foreseeable that the interplay of concepts from laser spectroscopy, quantum optics, photophysics, photochemistry, nanotechnology, and biophysics will open up many new avenues for optical imaging. Advances in all these research areas will ultimately confront the signal-to-noise barrier. New developments for efficient collection of photons and their implementation in lab-on-a-chip solutions, progress in synthesizing new labels, as well as better detectors and laser sources will all play a part in pushing this final frontier. In this dissertation, we have demonstrated that optical microscopy is a powerful tool. A tool that will someday help us to visualize and explore matter, be it biological or inorganic, at arbitrary resolution. It will let us see this *new visible World* in a light no one has seen it before.

Bibliography

[1] D. Bardell, "The Invention of the Microscope," BIOS **75**, 78 (2004).

[2] P. Török and F.-J. Kao, *Optical Imaging and Microscopy* (Springer, New York, 2007).

[3] E. Abbe, "Beiträge zur Theorie des Mikroskops und der mikroskopischen Wahrnehmung," Arch. Mikrosk. Anat. **IX**, 413 (1873).

[4] A. V. Narlikar and Y. Fu, *Oxford Handbook of Nanoscience and Technology* (Oxford University Press, New York, 2010).

[5] C. J. R. Sheppard, J. N. Gannaway, D. Walsh, and T. Wilson, "Scanning Optical Microscope for the Inspection of Electronic Devices," Microcircuit Engineering Conference, Cambridge (1978).

[6] D. Pohl, W. Denk, and M. Lanz, "Optical stethoscopy: Image recording with resolution $\lambda/20$," Appl. Phys. Lett. **44**, 651 (1984).

[7] M. G. L. Gustafsson, "Surpassing the lateral resolution limit by a factor of two using structured illumination microscopy," J. Microsc. **198**, 82 (2000).

[8] C. Hettich, C. Schmitt, J. Zitzmann, S. Kühn, I. Gerhardt, and V. Sandoghdar, "Nanometer resolution and coherent optical dipole coupling of two individual molecules." Science **298**, 385 (2002).

[9] X. Qu, D. Wu, L. Mets, and N. F. Scherer, "Nanometer-localized multiple single-moelcule fluorescence microscopy," Proc. Natl. Acad. Sci. USA **101**, 11,298 (2004).

[10] M. P. Gordon, T. Ha, and P. R. Selvin, "Single-molecule high-resolution imaging with photobleaching," Proc. Natl. Acad. Sci. USA **101**, 6462 (2004).

[11] G. Donnert, J. Keller, R. Medda, M. A. Andrei, S. O. Rizzoli, R. Lührmann, R. Jahn, C. Eggeling, and S. W. Hell, "Macromolecular-scale resolution in biological fluorescence microscopy," Proc. Natl. Acad. Sci. USA **103**, 11,440 (2006).

[12] S. Bretschneider, C. Eggeling, and S. W. Hell, "Breaking the Diffraction Barrier in Fluorescence Microscopy by Optical Shelving," Phys. Rev. Lett. **98**, 218,103 (2007).

[13] A. Pertsinidis, Y. Zhang, and S. Chu, "Subnanometre single-molecule localization, registration and distance measurements." Nature **466**, 647 (2010).

[14] J. C. Vaughan, S. Jia, and X. Zhuang, "Ultrabright photoactivatable fluorophores created by reductive caging," Nat. Methods **9**, 1181 (2012).

[15] D. Wildanger, B. R. Patton, H. Schill, L. Marseglia, J. P. Hadden, S. Knauer, A. Schönle, J. G. Rarity, J. L. O'Brien, S. W. Hell, and J. M. Smith, "Solid Immersion Facilitates Fluorescence Microscopy with Nanometer Resolution and Sub-Angström Emitter Localization," Adv. Mater. **24**, 309 (2012).

[16] T. Basché, W. E. Moerner, M. Orrit, and U. P. Wild, *Single-Molecule Optical Detection, Imaging and Spectroscopy* (Wiley-VCH, Weinheim, 1997).

[17] A. Ishijima and T. Yanagida, "Single molecule nanobioscience," Trends Biochem. Sci. **26**, 438 (2001).

[18] F. Kulzer and M. Orrit, "Single-Molecule Optics," Annu. Rev. Phys. Chem. **55**, 585 (2004).

[19] M. F. Juette, D. S. Terry, M. R. Wasserman, Z. Zhou, R. B. Altman, Q. Zheng, and S. C. Blanchard, "The bright future of single-molecule fluorescence imaging," Curr. Opin. Chem. Biol. **20**, 103 (2014).

[20] W. F. Talbot, "Experiments on light," Phil. Mag. **5**, 321 (1834).

[21] F. Zernike, "How I Discovered Phase Contrast," Science **121**, 345 (1955).

[22] J. B. Pawley, *Handbook of Biological Confocal Microscopy* (Springer, New York, 2006).

[23] A. Coons and M. Kaplan, "Localization of antigen in tissue cells; improvements in a method for the detection of antigen by means of fluorescent antibody," J. Exp. Med. **91**, 1 (1950).

[24] R. Y. Tsien and A. Waggoner, "Fluorophores for confocal microscopy," in *Handbook of biological confocal microscopy*, J. B. Pawley, ed. (Springer, New York, 2006).

[25] R. Y. Tsien, "The Green Fluorescent Protein," Annu. Rev. Biochem. **67**, 509 (1998).

[26] W. Denk, J. H. Strickler, and W. W. Webb, "Two-photon laser scanning fluorescence microscopy," Science **248**, 73 (1990).

[27] M. Delhaye and P. Dhamelincourt, "Raman microprobe and microscope with laser excitation," J. Raman Spectrosc. **3**, 33 (1975).

[28] P. D. Maker and R. W. Terhune, "Study of Optical Effects Due to an Induced Polarization Third Order in the Electric Field Strength," Phys. Rev. **137**, A801 (1965).

[29] C. W. Freudiger, W. Min, B. G. Saar, S. Lu, G. R. Holtom, C. He, J. C. Tsai, J. X. Kang, and X. S. Xie, "Label-Free Biomedical Imaging with High Sensitivity by Stimulated Raman Scattering Microscopy," Science **322**, 1857 (2008).

[30] S. Fine and W. P. Hansen, "Optical Second Harmonic Generation in Biological Systems," Appl. Opt. **10**, 2350 (1971).

[31] I. Freund, M. Deutsch, and A. Sprecher, "Connective tissue polarity. Optical second-harmonic microscopy, crossed-beam summation, and small-angle scattering in rat-tail tendon." Biophys. J. **50**, 693 (1986).

[32] G. Nomarski, "Microinterféromètre différential à ondes polarisées," J. Phys. Radium **16**, 9S (1955).

[33] R. Hoffman and L. Gross, "Modulation Contrast," Appl. Opt. Microscope **14**, 1169 (1975).

[34] K. Lindfors, T. Kalkbrenner, P. Stoller, and V. Sandoghdar, "Detection and spectroscopy of gold nanoparticles using supercontinuum white light confocal microscopy," Phys. Rev. Lett. **93**, 037,401 (2004).

[35] P. Kukura, H. Ewers, C. Müller, A. Renn, A. Helenius, and V. Sandoghdar, "High-speed nanoscopic tracking of the position and orientation of a single virus," Nat. Methods **6**, 923 (2009).

[36] J. Jung, S. Weisenburger, S. Albert, D. F. Gilbert, O. Friedrich, V. Eulenburg, J. Kornhuber, and T. W. Groemer, "Performance of scientific cameras with different sensor types in measuring dynamic processes in fluorescence microscopy," Microsc. Res. Tech. **76**, 835 (2013).

[37] F. Rieke and D. A. Baylor, "Single-photon detection by rod cells of the retina," Rev. Mod. Phys. **70**, 1027 (1998).

[38] S. Cova, M. Bertolaccini, and C. Bussolati, "The measurement of luminescence waveforms by single photon techniques," Phys. Status. Solid. A **18**, 11 (1973).

[39] M. Orrit and J. Bernard, "Single pentacene molecules detected by fluorescence excitation in a p-terphenyl crystal," Phys. Rev. Lett. **65**, 2716 (1990).

[40] E. Betzig and R. J. Chichester, "Single Molecules Observed by Near-Field Scanning Optical Microscopy," Science **262**, 1422 (1993).

[41] S. Nie, D. T. Chiu, and R. N. Zare, "Probing individual molecules with confocal fluorescence microscopy," Science **266**, 1018 (1994).

[42] S. Nie and S. R. Emory, "Probing Single Molecules and Single Nanoparticles by Surface-Enhanced Raman Scattering," Science **275**, 1102 (1997).

[43] P. L. Stiles, J. A. Dieringer, N. C. Shah, and R. P. V. Duyne, "Surface-Enhanced Raman Spectroscopy," Annu. Rev. Anal. Chem. **1**, 601 (2008).

[44] V. Jacobsen, P. Stoller, C. Brunner, V. Vogel, and V. Sandoghdar, "Interferometric optical detection and tracking of very small gold nanoparticles at a water-glass interface," Opt. Express **14**, 405 (2006).

[45] P. Kukura, M. Celebrano, A. Renn, and V. Sandoghdar, "Imaging a Single Quantum Dot When It Is Dark," Nano Lett. **9**, 926 (2009).

[46] P. Kukura, M. Celebrano, A. Renn, and V. Sandoghdar, "Single-Molecule Sensitivity in Optical Absorption at Room Temperature," J. Phys. Chem. Lett. **1**, 3323 (2010).

[47] M. Celebrano, P. Kukura, A. Renn, and V. Sandoghdar, "Single molecule imaging by optical absorption," Nat. Photonics **5**, 95 (2011).

[48] M. Piliarik and V. Sandoghdar, "Direct optical sensing of single unlabeled small proteins and super-resolution microscopy of their binding sites," Nat. Commun. **5**, 4495 (2014).

[49] H. von Helmholtz, "Die theoretische Grenze für die Leistungsfähigkeit der Mikroskope," Ann. Phys. (Berlin) p. 557 (1874).

[50] J. W. Goodman, *Introduction to Fourier Optics* (McGraw-Hill, New York, 1996).

[51] J. W. Rayleigh, "On the theory of optical images, with special reference to the microscope," Philos. Mag. Ser. **42**, 167 (1896).

[52] G. B. Airy, "On the diffraction of an object-glass with circular aperture," Trans. Cambridge Philos. Soc. **5**, 283 (1835).

[53] L. Rayleigh, "Investigations in optics with special reference to the spectroscope," Philos. Mag. **8**, 261–274, 403–411, 477–486 (1879).

[54] R. Dorn, S. Quabis, and G. Leuchs, "Sharper Focus for a Radially Polarized Light Beam," Phys. Rev. Lett. **91**, 233,901 (2003).

[55] M. Born and E. Wolf, *Principles of Optics: Electromagnetic Theory of Propagation, Interference and Diffraction of Light* (Cambridge University Press, 1997).

[56] F. Träger, *Handbook of Lasers and Optics* (Springer, New York, 2007).

[57] M. Minsky, "Microscopy apparatus," (1961). US Patent 3,013,467, URL `http://www.google.de/patents/US3013467`.

[58] C. J. R. Sheppard and A. Choudhury, "Image formation in the scanning microscope," Optica Acta **24**, 1051 (1977).

[59] C. J. R. Sheppard, "Super-resolution in confocal imaging," Optik **80**, 53 (1988).

[60] C. B. Müller and J. Enderlein, "Image scanning microscopy," Phys. Rev. Lett. **104**, 198,101 (2010).

[61] M. G. L. Gustafsson, D. A. Agard, and J. W. Sedat, "I^5M: 3D widefield light microscopy with better than 100 nm axial resolution," J. Microsc. **195**, 10 (1999).

[62] L. Shao, B. Isaac, S. Uzawa, D. A. Agard, J. W. Sedat, and M. G. L. Gustafsson, "I^5S: Wide-field light microscopy with 100-nm-scale resolution in three dimensions," Biophys. J. **94**, 4971 (2008).

[63] B. C. Chen, W. R. Legant, K. Wang, L. Shao, D. E. Milkie, M. W. Davidson, C. Janetopoulos, X. S. Wu, J. A. H. 3rd, Z. Liu, B. P. English, Y. Mimori-Kiyosue, D. P. Romero, A. T. Ritter, J. Lippincott-Schwartz, L. Fritz-Laylin, R. D. Mullins, D. M. Mitchell, J. N. Bembenek, A. C. Reymann, R. Böhme, S. W. Grill, J. T. Wang, G. Seydoux, U. S. Tulu, D. P. Kiehart, and E. Betzig, "Lattice light-sheet microscopy: Imaging molecules to embryos at high spatiotemporal resolution," Science **346**, 1257,998 (2014).

[64] K. Wicker and R. Heintzmann, "Resolving a misconception about structured illumination," Nat. Photonics **8**, 342 (2014).

[65] E. H. Synge, "A suggested method for extending microscopic resolution into the ultra-microscopic region," Philos. Mag. **6**, 356 (1928).

[66] V. G. Veselago, "The electrodynamics of substances with simultaneously negative values of permittivity and permeability," Sov. Phys. Usp. **10**, 509 (1968).

[67] J. Valentine, S. Zhang, T. Zentgraf, E. Ulin-Avila, D. A. Genov, G. Bartal, and X. Zhang, "Three-dimensional optical metamaterial with a negative refractive index," Nature **455**, 376 (2008).

[68] J. B. Pendry, "Negative Refraction Makes a Perfect Lens," Phys. Rev. Lett. **85**, 3966 (2000).

[69] N. Fang, H. Lee, C. Sun, and X. Zhang, "Sub–Diffraction-Limited Optical Imaging with a Silver Superlens," Science **308**, 534 (2005).

[70] Z. Liu, H. Lee, Y. Xiong, C. Sun, and X. Zhang, "Far-Field Optical Hyperlens Magnifying Sub-Diffraction-Limited Objects," Science **315**, 1686 (2007).

[71] J. Rho, Z. Ye, Y. Xiong, X. Yin, Z. Liu, H. Choi, G. Bartal, and X. Zhang, "Spherical hyperlens for two-dimensional sub-diffractional imaging at visible frequencies," Nat. Commun. **1**, 143 (2010).

[72] S. W. Hell, "Biographical," in *The Nobel Prizes 2014*, T. N. Foundation, ed. (Watson Publishing International LLC, Sagamore Beach, 2015).

[73] S. W. Hell, "Double-confocal scanning microscope," (1996). EP Patent 0,491,289, URL `http://www.google.de/patents/EP0491289B1`.

[74] S. Hell, E. Stelzer, S. Lindek, and C. Cremer, "Confocal microscopy with an increased detection aperture: type-B 4Pi confocal microscopy," Optics Letters **19**, 222 (1994).

[75] S. W. Hell and J. Wichmann, "Breaking the diffraction resolution limit by stimulated emission: stimulated-emission-depletion fluorescence microscopy," Opt. Lett. **19**, 780 (1994).

[76] T. A. Klar and S. W. Hell, "Subdiffraction resolution in far-field fluorescence microscopy," Opt. Lett. **24**, 954 (1999).

[77] F. Göttfert, C. A. Wurm, V. Mueller, S. Berning, V. C. Cordes, A. Honigmann, and S. W. Hell, "Coaligned Dual-Channel STED Nanoscopy and Molecular Diffusion Analysis at 20 nm Resolution," Biophys. J. **105**, L01 (2013).

[78] L. Meyer, D. Wildanger, R. Medda, A. Punge, S. O. Rizzoli, G. Donnert, and S. W. Hell, "Dual-Color STED Microscopy at 30-nm Focal-Plane Resolution," Small **4**, 1095 (2008).

[79] K. I. Willig, R. R. Kellner, R. Medda, B. Hein, S. Jakobs, and S. W. Hell, "Nanoscale resolution in GFP-based microscopy," Nat. Methods **3**, 721 (2006).

[80] S. Arroyo-Camejo, M. P. Adam, M. Besbes, J. P. Hugonin, V. Jacques, J. J. Greffet, J. F. Roch, S. W. Hell, and F. Treussart, "Stimulated Emission Depletion Microscopy Resolves Individual Nitrogen Vacancy Centers in Diamond Nanocrystals," ACS Nano **7**, 10,912 (2013).

[81] B. Yang, J.-B. Trebbia, R. Baby, P. Tamarat, and B. Lounis, "Optical Nanoscopy with Excited State Saturation at Liquid Helium Temperatures," Nat. Photonics **9**, 658 (2015).

[82] W. R. Silva, C. T. Graefe, and R. R. Frontiera, "Toward Label-Free Super-Resolution Microscopy," ACS Photonics **3**, 79 (2016).

[83] S. W. Hell and M. Kroug, "Ground-state-depletion fluorescence microscopy: A concept for breaking the diffraction resolution limit," Appl. Phys. B **5**, 495 (1995).

[84] M. Hofmann, C. Eggeling, S. Jakobs, and S. W. Hell, "Breaking the diffraction barrier in fluorescence microscopy at low light intensities by using reversibly photoswitchable proteins," Proc. Natl. Acad. Sci. USA **102**, 17,565 (2005).

[85] S. W. Hell, "Far-Field Optical Nanoscopy," Science **316**, 1153 (2007).

[86] R. Heintzmann, T. Jovin, and C. Cremer, "Saturated patterned excitation microscopy – a concept for optical resolution improvement," J. Opt. Soc. Am. A **19**, 1599 (2002).

[87] M. G. L. Gustafsson, "Nonlinear structured-illumination microscopy: Wide-field fluorescence imaging with theoretically unlimited resolution," Proc. Natl. Acad. Sci. USA **102**, 13,081 (2005).

[88] P. Bingen, M. Reuss, J. Engelhardt, and S. W. Hell, "Parallelized STED fluorescence nanoscopy," Opt. Express **19**, 23,716 (2011).

[89] A. Chmyrov, J. Keller, T. Grotjohann, M. Ratz, E. d'Este, S. Jakobs, C. Eggeling, and S. W. Hell, "Nanoscopy with more than 100,000 'doughnuts'," Nat. Methods **10**, 737 (2013).

[90] B. Yang, F. Przybilla, M. Mestre, J.-B. Trebbia, and B. Lounis, "Parallelized STED fluorescence nanoscopy," Opt. Express **22**, 5581 (2014).

[91] J. Hotta, E. Fron, P. Dedecker, K. P. F. Janssen, C. Li, K. Müllen, B. Harke, J. Bückers, S. W. Hell, and J. Hofkens, "Spectroscopic Rationale for Efficient Stimulated-Emission Depletion Microscopy Fluorophores," J. Am. Chem. Soc. **132**, 5021 (2010).

[92] Abberior GmbH, "Overview of Abberior fluorescence markers," `http://www.abberior.com/references/dye-overview/overview-abberior-dyes/` (2014).

[93] J. G. Danzl, S. C. Sidenstein, C. Gregor, N. T. Urban, P. Ilgen, S. Jakobs, and S. W. Hell, "Coordinate-targeted fluorescence nanoscopy with multiple off states," Nat. Photonics **10**, 122 (2016).

[94] G. Donnert, J. Keller, C. A. Wurm, S. O. Rizzoli, V. Westphal, A. Schönle, R. Jahn, S. Jakobs, C. Eggeling, and S. W. Hell, "Two-Color Far-field Fluorescence Nanoscopy," Biophys. J. **92**, L67 (2007).

[95] J. Tønnesen, F. Nadrigny, K. I. Willig, R. Wedlich-Söldner, and U. V. Nägerl, "Two-Color STED Microscopy of Living Synapses Using A Single Laser-Beam Pair," Biophys. J. **101**, 2545 (2011).

[96] J. Bückers, D. Wildanger, G. Vicidomini, L. Kastrup, and S. W. Hell, "Simultaneous multi-lifetime multi-color STED imaging for colocalization analyses," Opt. Express **19**, 3130 (2011).

[97] T. J. Gould, D. Burke, J. Bewersdorf, and M. J. Booth, "Adaptive optics enables 3D STED microscopy in aberrating specimens," Opt. Express **20**, 20,998 (2012).

[98] M. Booth, D. Andrade, D. Burke, B. Patton, and M. Zurauskas, "Aberrations and adaptive optics in super-resolution microscopy," Microscopy **64**, 251 (2015).

[99] S. Völker, "Hole-Burning Spectroscopy," Annu. Rev. Phys. Chem. **40**, 499 (1989).

[100] T. Hirschfeld, "Optical Microscopic Observation of Single Small Molecules," Appl. Opt. **15**, 2965 (1976).

[101] W. E. Moerner and L. Kador, "Optical detection and spectroscopy of single molecules in a solid," Phys. Rev. Lett. **62**, 2535 (1989).

[102] W. P. Ambrose and W. E. Moerner, "Fluorescence spectroscopy and spectral diffusion of single impurity molecules in a crystal," Nature **349**, 225 (1991).

[103] J. K. Trautman, J. J. Macklin, L. E. Brus, and E. Betzig, "Near-field spectroscopy of single molecules at room temperature," Nature **369**, 40 (1994).

[104] W. P. Ambrose, P. M. Goodwin, R. A. Keller, and J. C. Martin, "Alterations of Single Molecule Fluorescence Lifetimes in Near-Field Optical Microscopy," Science **265**, 364 (1994).

[105] R. M. Dickson, A. B. Cubitt, R. Y. Tsien, and W. E. Moerner, "On/off blinking and switching behaviour of single molecules of green fluorescent protein," Nature **388**, 355 (1997).

[106] M. Kuno, D. P. Fromm, H. F. Hamann, A. Gallagher, and D. J. Nesbitt, ""On"/"off" fluorescence intermittency of single semiconductor quantum dots," J. Chem. Phys **115**, 1028 (2001).

[107] K. T. Shimizu, R. G. Neuhauser, C. A. Leatherdale, S. A. Empedocles, W. K. Woo, and M. G. Bawendi, "Blinking statistics in single semiconductor nanocrystal quantum dots," Phys. Rev. B **63**, 205,311–205,316 (2001).

[108] R. Zondervan, F. Kulzer, S. B. Orlinskii, and M. Orrit, "Photoblinking of rhodamine 6G in poly(vinyl alcohol): radical dark state formed through the triplet," J. Phys. Chem. A **107**, 6770 (2003).

[109] K. Suzuki, S. Habuchi, and M. Vacha, "Blinking of single dye molecules in a polymer matrix is correlated with free volukme in polymers," Chem. Phys. Lett. **505**, 157 (2011).

[110] J. P. Hoogenboom, J. Hernando, E. M. H. P. van Dijk, N. F. van Hulst, and M. F. García-Parajó, "Power-Law Blinking in the Fluorescence of Single Organic Molecules," ChemPhysChem **8**, 823 (2007).

[111] J. Schuster, F. Cichos, and C. von Borczyskowski, "Influence of self-trapped states on the fluorescence intermittency of single molecules," Appl. Phys. Lett. **87**, 051,915 (2005).

[112] J. P. Hoogenboom, E. M. H. P. van Dijk, J. Hernando, N. F. van Hulst, and M. F. García-Parajó, "Power-Law-Distributed Dark States are the Main Pathway for Photobleaching of Single Organic Molecules," Phys. Rev. Lett. **95**, 097,401 (2005).

[113] E. K. L. Yeow, S. M. Melnikov, T. D. M. Bell, F. C. De Schryver, and J. Hofkens, "Characterizing the fluorescence intermittency and photobleaching kinetics of dye molecules immobilized on a glass surface." J. Phys. Chem. A **110**, 1726–34 (2006).

[114] S. V. Orlov, A. V. Naumov, Y. G. Vainer, and L. Kador, "Spectrally resolved analysis of fluorescence blinking of single dye molecules in polymers at low temperatures," J. Chem. Phys. **137**, 194,903 (2012).

[115] D. Sluss, C. Bingham, M. Burr, E. D. Bott, E. A. Riley, and P. J. Reid, "Temperature-dependent fluorescence intermittency for single molecules of violamine R in poly(vinyl alocohol)," J. Mater. Chem. **19**, 7561 (2009).

[116] R. J. Pfab, J. Zimmermann, C. Hettich, I. Gerhardt, A. Renn, and V. Sandoghdar, "Aligned terrylene molecules in a spin-coated ultrathin crystalline film of p-terphenyl," Chem. Phys. Lett. **387**, 490 (2004).

[117] W. K. Heisenberg, *The Physical Principles of the Quantum Theory* (Chicago University Press, 1930).

[118] J. Gelles, B. J. Schnapp, and M. P. Sheetz, "Tracking kinesin-driven movements with nanometre-scale precision," Nature **331**, 450 (1988).

[119] R. E. Thompson, D. R. Larson, and W. W. Webb, "Precise nanometer localization analysis for individual fluorescent probes." Biophys. J. **82**, 2775 (2002).

[120] R. J. Ober, S. Ram, and E. S. Ward, "Localization Accuracy in Single-Molecule Microscopy," Biophys J. **86**, 1185 (2004).

[121] K. I. Mortensen, L. S. Churchman, J. A. Spudich, and H. Flyvbjerg, "Optimized localization analysis for single-molecule tracking and super-resolution microscopy." Nat. Methods **7**, 377 (2010).

[122] S. A. McKinney, C. S. Murphy, K. L. Hazelwood, M. W. Davidson, and L. L. Looger, "A bright and photostable photoconvertible fluorescent protein." Nat. Methods **6**, 131 (2009).

[123] A. Yildiz, J. N. Forkey, S. A. McKinney, T. Ha, Y. E. Goldman, and P. R. Selvin, "Myosin V walks hand-over-hand: single fluorophore imaging with 1.5-nm localization." Science **300**, 2061–5 (2003).

[124] S. F. Lee, Q. Vèrolet, and A. Fürstenberg, "Improved Super-Resolution Microscopy with Oxazine Fluorophores in Heavy Water," Angew. Chem. Int. Ed. **52**, 8948–8951 (2013).

[125] M. J. Saxton and K. Jacobson, "Single-particle tracking: applications to membrane dynamics," Annu. Rev. Biophys. Biomol. Struct. **26**, 373 (1997).

[126] M. P. Sheetz and J. A. Spudich, "Movement of myosin-coated fluorescent beads on actin cables in vitro," Nature **303**, 31 (1983).

[127] G. J. Schütz, H. Schindler, and T. Schmidt, "Single-molecule microscopy on model membranes reveals anomalous diffusion," Biophys. J. **73**, 1073 (1997).

[128] T. Fujiwara, K. Ritchie, H. Murakoshi, K. Jacobson, and A. Kusumi, "Phospholipids undergo hop diffusion in compartmentalized cell membrane," J. Cell Biol. **157**, 1071 (2002).

[129] A. Kusumi, C. Nakada, K. Ritchie, K. Murase, K. Suzuki, H. Murakoshi, R. S. Kasai, J. Kondo, and T. Fujiwara, "Paradigm shift of the plasma membrane concept from the two-dimensional continuum fluid to the partitioned fluid: high-speed single-molecule tracking of membrane molecules," Annu. Rev. Biophys. Biomol. Struct. **34**, 351 (2005).

[130] D. Lasne, G. A. Blab, S. Berciaud, M. Heine, L. Groc, D. Choquet, L. Cognet, and B. Lounis, "Single nanoparticle photothermal tracking (SNaPT) of 5 nm gold beads in live cells," Biophys. J. **91**, 4598 (2006).

[131] C.-L. Hsieh, S. Spindler, J. Ehrig, and V. Sandoghdar, "Tracking Single Particles on Supported Lipid Membranes: Multimobility Diffusion and Nanoscopic Confinement," J. Phys. Chem. B **118**, 1545 (2014).

[132] A. V. Gorshkov, L. Jiang, M. Greiner, P. Zoller, and M. D. Lukin, "Coherent Quantum Optical Control with Subwavelength Resolution," Phys. Rev. Lett. **100**, 093,005 (2008).

[133] W. S. Bakr, J. I. Gillen, A. Peng, S. Fölling, and M. Greiner, "A quantum gas microscope for detecting single atoms in a Hubbard-regime optical lattice," Nature **462**, 74 (2009).

[134] J. F. Sherson, C. Weitenberg, M. Endres, M. Cheneau, I. Bloch, and S. Kuhr, "Single-atom-resolved fluorescence imaging of an atomic Mott insulator," Nature **467**, 68 (2010).

[135] E. Betzig, "Proposed method for molecular optical imaging." Opt. Lett. **20**, 237 (1995).

[136] F. Güttler, T. Irngartinger, T. Plakhotnik, A. Renn, and U. P. Wild, "Fluorescence microscopy of single molecules," Chem. Phys. Lett. **217**, 393 (1994).

[137] T. Utikal, E. Eichhammer, L. Petersen, A. Renn, S. Götzinger, and V. Sandoghdar, "Spectroscopic detection and state preparation of a single praseodymium ion in a crystal," Nat. Commun. **5**, 3627 (2014).

[138] T. D. Lacoste, X. Michalet, F. Pinaud, D. S. Chemla, A. P. Alivisatos, and S. Weiss, "Ultrahigh-resolution multicolor colocalization of single fluorescent probes," Proc. Natl. Acad. Sci. USA **97**, 9461 (2000).

[139] E. Betzig, G. H. Patterson, R. Sougrat, O. W. Lindwasser, S. Olenych, J. S. Bonifacino, M. W. Davidson, J. Lippincott-Schwartz, and H. F. Hess, "Imaging intracellular fluorescent proteins at nanometer resolution." Science **313**, 1642 (2006).

[140] S. T. Hess, T. Girirajan, and M. Mason, "Ultra-High Resolution Imaging by Fluorescence Photoactivation Localization Microscopy," Biophys. J. **91**, 4258–4272 (2006).

[141] M. J. Rust, M. Bates, and X. Zhuang, "Sub-diffraction-limit imaging by stochastic optical reconstruction microscopy (STORM)," Nat. Methods **3**, 793 (2006).

[142] J. Ries, C. Kaplan, E. Platonova, H. Eghlidi, and H. Ewers, "A simple, versatile method for GFP-based super-resolution microscopy via nanobodies," Nat. Methods **9**, 582 (2012).

[143] F. Opazo, M. Levy, M. Byrom, C. Schafer, C. Geisler, T. W. Groemer, A. D. Ellington, and S. O. Rizzoli, "Aptamers as potential tools for super-resolution microscopy," Nat. Methods **9**, 938 (2012).

[144] B. R. Terry, E. K. Matthews, and J. Haseloff, "Molecular characterisation of recombinant green fluorescent protein by fluorescence correlation microscopy," Biochem. Bioph. Res. Co. **217**, 21 (1995).

[145] A. Sharonov and R. M. Hochstrasser, "Wide-field subdiffraction imaging by accumulated binding of diffusing probes," Proc. Natl. Acad. Sci. USA **103**, 18,911 (2006).

[146] R. Jungmann, C. Steinhauer, M. Scheible, A. Kuzyk, P. Tinnefeld, and F. C. Simmel, "Single-molecule kinetics and super-resolution microscopy by fluorescence imaging of transient binding on DNA origami," Nano Lett. **10**, 4756 (2010).

[147] I. Schoen, J. Ries, E. Klotzsch, H. Ewers, and V. Vogel, "Binding-Activated Localization Microscopy of DNA Structures," Nano Lett. **11**, 4008 (2011).

[148] S. van de Linde, S. Wolter, M. Heilemann, and M. Sauer, "The effect of photoswitching kinetics and labeling densities on super-resolution fluorescence imaging," J. Biotechnol. **149**, 260 (2010).

[149] D. Bourgeois and V. Adam, "Reversible photoswitching in fluorescent proteins: A mechanistic view," IUBMB Life **64**, 482 (2012).

[150] J. Vogelsang, R. Kasper, C. Steinhauer, B. Person, M. Heilemann, M. Sauer, and P. Tinnefeld, "A Reducing and Oxidizing System Minimizes Photobleaching and Blinking of Fluorescent Dyes," Angew. Chem. Int. Ed. **47**, 5465 (2008).

[151] T. Ha and P. Tinnefeld, "Photophysics of Fluorescent Probes for Single-Molecule Biophysics and Super-Resolution Imaging," Annu. Rev. Phys. Chem. **63**, 595–617 (2012).

[152] Y.-W. Chang, S. Chen, E. I. Tocheva, A. Treuner-Lange, S. Löbach, L. Søgaard-Andersen, and G. J. Jensen, "Correlated cryogenic photoactivated localization microscopy and cryo-electron tomography," Nat. Methods **11**, 737 (2014).

[153] R. Kaufmann, P. Schellenberger, E. Seiradake, I. M. Dobbie, E. Y. Jones, I. Davis, C. Hagen, and K. Grünewald, "Super-Resolution Microscopy Using Standard Fluorescent Proteins in Intact Cells under Cryo-Conditions," Nano Lett. **14**, 4171 (2014).

[154] R. Zondervan, F. Kulzer, H. van der Meer, J. Disselhorst, and M. Orrit, "Laser-driven microsecond temperature cycles analyzed by fluorescence polarization microscopy," Biophys. J. **90**, 2958 (2006).

[155] A. P. Bartko and R. M. Dickson, "Imaging Three-Dimensional Single Molecule Orientations," J. Phys. Chem. B **103**, 11,237 (1999).

[156] J. Enderlein, E. Toprak, and P. R. Selvin, "Polarization effect on position accuracy of fluorophore localization." Opt. Express **14**, 8111 (2006).

[157] F. Aguet, S. Geissbühler, I. Märki, T. Lasser, and M. Unser, "Super-resolution orientation estimation and localization of fluorescent dipoles using 3-D steerable filters," Opt. Express **17**, 6829–6848 (2009).

[158] S. Stallinga and B. Rieger, "Position and orientation estimation of fixed dipole emitters using an effective Hermite point spread function model," Opt. Express **20**, 5896 (2012).

[159] C. E. Shannon, "Communication in the presence of noise," Proc. IRE **37**, 10 (1949).

[160] T. Förster, "Zwischenmolekulare Energiewanderung und Fluoreszenz," Ann. Phys. (Berlin) **437**, 55 (1948).

[161] R. Luchoswki, E. G. Matveeva, I. Gryczynski, E. A. Terpetschnig, L. Patsenker, G. Laczko, J. Borejdo, and Z. Gryczynski, "Single Molecule Studies of Multiple-Fluorophore Labeled Antibodies. Effect of Homo-FRET on the Number of Photons Available Before Photobleaching," Curr. Pharm. Biotechnol. **9**, 411 (2008).

[162] S. J. Holden, S. Uphoff, and A. N. Kapanidis, "DAOSTORM: an algorithm for high-density super-resolution microscopy," Nat. Methods **8**, 279 (2011).

[163] F. Huang, S. L. Schwartz, J. M. Byars, and K. A. Lidke, "Simultaneous multiple-emitter fitting for single molecule super-resolution imaging," Biomed. Opt. Express **2**, 1377 (2011).

[164] B. Huang, W. Wang, M. Bates, and X. Zhuang, "Three-dimensional super-resolution imaging by stochastic optical reconstruction microscopy." Science **319**, 810 (2008).

[165] B. Huang, S. A. Jones, B. Brandenburg, and X. Zhuang, "Whole-cell 3D STORM reveals interactions between cellular structures with nanometer-scale resolution," Nat. Methods **5**, 1047 (2008).

[166] S. Jia, J. C. Vaughan, and X. Zhuang, "Isotropic three-dimensional super-resolution imaging with a self-bending point spread function," Nat. Photonics **8**, 302 (2014).

[167] M. F. Juette, T. J. Gould, M. D. Lessard, M. J. Mlodzianoski, B. S. Nagpure, and B. T. Bennett, "Three-dimensional sub-100 nm resolution fluorescence microscopy of thick samples," Nat. Methods **5**, 527 (2008).

[168] S. Abrahamsson, J. Chen, B. Hajj, S. Stallinga, A. Y. Katsov, J. Wisniewski, G. Mizuguchi, P. Soule, F. Mueller, C. D. Darzacq, X. Darzacq, C. Wu, C. I. Bargmann, D. A. Agard, M. Dahan, and M. G. L. Gustafsson, "Fast multicolor 3D imaging using aberration-corrected multifocus microscopy," Nat. Methods **10**, 60 (2013).

[169] S. R. P. Pavani, M. A. Thompson, J. S. Biteen, S. J. Lord, N. Liu, R. J. Twieg, R. Piestun, and W. E. Moerner, "Three-dimensional, single-molecule fluorescence imaging beyond the diffraction limit by using a double-helix point spread function," Proc. Natl. Acad. Sci. USA **106**, 2995 (2009).

[170] Y. Cheng, N. Grigorieff, P. A. Penczek, and T. Walz, "A Primer to Single-Particle Cryo-Electron Microscopy," Cell **161**, 438 (2015).

[171] M. Bates, B. Huang, G. T. Dempsey, and X. Zhuang, "Multicolor Super-Resolution Imaging with Photo-Switchable Fluorescent Probes," Science **317**, 1749 (2007).

[172] S. van de Linde, U. Endesfelder, A. Mukherjee, M. Schüttpelz, G. Wiebusch, S. Wolter, M. Heilemann, and M. Sauer, "Multicolor photoswitching microscopy for subdiffraction-resolution fluorescence imaging," Photochem. Photobiol. Sci. **8**, 465 (2009).

[173] A. Lampe, V. Haucke, S. J. Sigrist, M. Heilemann, and J. Schmoranzer, "Multi-colour direct STORM with red emitting carbocyanines," Biol. Cell **104**, 229 (2012).

[174] M. Bossi, J. Fölling, V. N. Belov, V. P. Boyarskiy, R. Medda, A. Egner, C. Eggeling, A. Schönle, and S. W. Hell, "Multicolor Far-Field Fluorescence Nanoscopy through Isolated Detection of Distinct Molecular Species," Nano Lett. **8**, 2463 (2008).

[175] R. Dixit and R. Cyr, "Cell damage and reactive oxygen species production induced by fluorescence microscopy: effect on mitosis and guidelines for non-invasive fluorescence microscopy," Plant J. **36**, 280 (2003).

[176] J. Huisken, J. Swoger, F. D. Bene, J. Wittbrodt, and E. H. Stelzer, "Optical sectioning deep inside live embryos by selective plane illumination microscopy," Science **305**, 1007 (2004).

[177] M. Mickoleit, B. Schmid, M. Weber, F. O. Fahrbach, S. Hombach, S. Reischauer, and J. Huisken, "High-resolution reconstruction of the beating zebrafish heart," Nat. Methods **11**, 919 (2014).

[178] T. A. Planchon, L. Gao, D. E. Milkie, M. W. Davidson, J. A. Galbraith, C. G. Galbraith, and E. Betzig, "Rapid three-dimensional isotropic imaging of living cells using Bessel beam plane illumination," Nat. Methods **8**, 417 (2011).

[179] U. Krzic, S. Gunther, T. E. Saunders, S. J. Streichan, and L. Hufnagel, "Multiview light-sheet microscope for rapid in toto imaging," Nat. Methods **9**, 730 (2012).

[180] R. Tomer, K. Khairy, F. Amat, and P. J. Keller, "Quantitative high-speed imaging of entire developing embryos with simultaneous multiview light-sheet microscopy," Nat. Methods **9**, 755 (2012).

[181] A. Sentenac, P. C. Chaumet, and K. Belkebir, "Beyond the Rayleigh criterion: grating assisted far-field optical diffraction tomography," Phys. Rev. Lett. **97**, 243,901 (2006).

[182] K. Belkebir, P. C. Chaumet, and A. Sentenac, "Influence of multiple scattering on three-dimensional imaging with optical diffraction tomography," J. Opt. Soc. Am. A **23**, 586 (2006).

[183] T. Zhang, Y. Ruan, G. Maire, D. Sentenac, A. Talneau, K. Belkebir, P. C. Chaumet, and A. Sentenac, "Full-polarized Tomographic Diffraction Microscopy Achieves a Resolution about One-Fourth of the Wavelength," Phys. Rev. Lett. **111**, 243,904 (2013).

[184] S. Ayas, G. Cinar, A. D. Ozkan, Z. Soran, O. Ekiz, D. Kocaay, A. Tomak, P. Toren, Y. Kaya, I. Tunc, H. Zareie, T. Tekinay, A. B. Tekinay, M. O. Guler, and A. Dana, "Label-Free Nanometer-Resolution Imaging of Biological Architectures through Surface Enhanced Raman Scattering," Sci. Rep. **3**, 2624 (2013).

[185] P. Wang, M. N. Slipchenko, J. Mitchell, C. Yang, E. O. Potma, X. Xu, and J.-X. Cheng, "Far-field imaging of non-fluorescent species with subdiffraction resolution," Nat. Photonics **7**, 449 (2013).

[186] O. Schwartz, J. M. Levitt, R. Tenne, S. Itzhakov, Z. Deutsch, and D. Oron, "Superresolution Microscopy with Quantum Emitters," Nano Lett. **13**, 5832 (2013).

[187] J.-M. Cui, F.-W. Sun, X.-D. Chen, Z.-J. Gong, and G.-C. Guo, "Quantum Statistical Imaging of Particles without Restriction of the Diffraction Limit," Phys. Rev. Lett. **110**, 153,901 (2013).

[188] C. Errico, J. Pierre, S. Pezet, Y. Desailly, Z. Lenkei, O. Couture, and M. Tanter, "Ultrafast ultrasound localization microscopy for deep super-resolution vascular imaging," Nature **527**, 499 (2015).

[189] S. Link and M. A. El-Sayed, "Shape and size dependence of radiative, non-radiative and photothermal properties of gold nanocrystals," Int. Rev. Phys. Chem. **19**, 409 (2000).

[190] H. Lee, T.-Y. Su, Y. Yonemaru, M.-Y. Lee, M. Yamanaka, K.-F. Huang, S. Kawata, K. Fujita, and S.-W. Chu, "Plasmon saturation induced super-resolution imaging," Proc. SPIE **8597**, 85,970P (2013).

[191] S.-W. Chu, H.-Y. Wu, Y.-T. Huang, T.-Y. Su, H. Lee, Y. Yonemaru, M. Yamanaka, R. Oketani, S. Kawata, S. Shoji, and K. Fujita, "Saturation and Reverse Saturation of Scattering in a Single Plasmonic Nanoparticle," ACS Photonics **1**, 32 (2014).

[192] S.-W. Chu, T.-Y. Su, R. Oketani, Y.-T. Huang, H.-Y. Wu, Y. Yonemaru, M. Yamanaka, H. Lee, G.-Y. Zhou, M.-Y. Lee, S. Kawata, and K. Fujita, "Measurement of a Saturated Emission of Optical Radiation from Gold Nanoparticles: Application to an Ultrahigh Resolution Microscope," Phys. Rev. Lett. **112**, 017,402 (2014).

[193] S. Watanabe, A. Punge, G. Hollopeter, K. I. Willig, R. J. Hobson, M. W. Davis, S. W. Hell, and E. M. Jorgensen, "Protein localization in electron micrographs using fluorescence nanoscopy," Nat. Methods **8**, 80 (2011).

[194] B. G. Kopek, G. Shtengel, C. S. Xu, D. A. Clayton, and H. F. Hess, "Correlative 3D super-resolution fluorescence and electron microscopy reveal the relationship of mitochondrial nucleoids to membranes," Proc. Natl. Acad. Sci. USA **109**, 6136 (2012).

[195] B. Harke, J. V. Chacko, H. Haschke, C. Canale, and A. Diaspro, "A novel nanoscopic tool by combining AFM with STED microscopy," Opt. Nanoscopy **1**, 3 (2012).

[196] J. V. Chacko, F. C. Zanacchi, and A. Diaspro, "Probing cytoskeletal structures by coupling optical superresolution and AFM techniques for a correlative approach," Cytoskeleton **70**, 729 (2013).

[197] A. Monserrate, S. Casado, and C. Flors, "Correlative Atomic Force Microscopy and Localization-Based Super-Resolution Microscopy: Revealing Labelling and Image Reconstruction Artefacts," ChemPhysChem **15**, 647 (2013).

[198] S. Wäldchen, J. Lehmann, T. Klein, S. van de Linde, and M. Sauer, "Light-induced cell damage in live-cell super-resolution microscopy," Sci. Rep. **5**, 15,348 (2015).

[199] D. Li, L. Shao, B.-C. Chen, X. Zhang, M. Zhang, B. Moses, D. E. Milkie, J. R. Beach, J. A. H. 3rd, M. Pasham, T. Kirchhausen, M. A. Baird, M. W. Davidson, P. Xu, and E. Betzig, "Extended-resolution structured illumination imaging of endocytic and cytoskeletal dynamics," Science **349**, 6251 (2015).

[200] Coherent, "Sapphire LP datasheet," `http://www.coherent.com/downloads/COHR_SapphireLP_DS_0514revG_5.pdf` (2015).

[201] Laser Quantum, "Finesse 532 datasheet," `http://www.laserquantum.com/download-ds.cfm?id=469` (2015).

[202] Toptica, "iBeam smart family datasheet," `http://www.toptica.com/fileadmin/user_upload/products/Diode_Lasers/Industrial_OEM/Single_Frequency/iBeam_smart_WS/toptica_BR_iBeam_smart_family.pdf` (2015).

[203] NKT Photonics, "SuperK Extreme datasheet," `http://www.nktphotonics.com/wp-content/uploads/2015/05/SuperK_EXTREME.pdf` (2015).

[204] Andor, "iXon 897 Performance sheet," X-4731 DV897DCS-BV (2006).

[205] Zeiss, "Objectives from Carl Zeiss," `http://www.zeiss.de/objectives` (2014).

[206] Mitutoyo, "Mitutoyo General Catalog," `http://www.mitutoyo.com/catalogs-brochures/mitutoyo-catalog-us-1003/` (2015).

[207] Thorlabs, "Product Catalog," `https://www.thorlabs.de/navigation.cfm` (2015).

[208] Semrock, "Individual Filters," `http://www.semrock.com/filters.aspx` (2015).

[209] J.-Y. Courtois, J.-M. Courty, and J. C. Mertz, "Internal dynamics of multilevel atoms near a vacuum-dielectric interface," Phys. Rev. A **53**, 1862 (1996).

[210] B. Jing, "Cryogenic Localization Microscopy," Master thesis, ETH Zürich, Switzerland (2011).

[211] J. Lakowicz, *Principles of Fluorescence Spectroscopy* (Springer, New York, 2006).

[212] I. Chen and A. Y. Ting, "Site-specific labeling of proteins with small molecules in live cells," Curr. Opin. Biotechnol. **16**, 35 (2005).

[213] S. Weiss, "Fluorescence Spectroscopy of Single Biomolecules," Science **283**, 1676 (1999).

[214] W. E. Moerner, "A Dozen Years of Single-Molecule Spectroscopy in Physics, Chemistry, and Biophysics," J. Phys. Chem. B **106**, 910 (2002).

[215] C. Joo, H. Balci, Y. Ishitsuka, C. Buranachai, and T. Ha, "Advances in single-molecule fluorescence methods for molecular biology," Annu. Rev. Biochem. **77**, 51 (2008).

[216] J. Frank, *Three-Dimensional Electron Microscopy of Macromolecular Assemblies* (Oxford University Press, New York, 2006).

[217] P. J. Hagerman, "Flexibility of DNA," Annu. Rev. Biophys. Biophys. Chem. **17**, 265 (1988).

[218] R. Verberk, A. M. van Oijen, and M. Orrit, "Simple model for the power-law blinking of single semiconductor nanocrystals," Phys. Rev. B **6**, 233,202 (2002).

[219] M. Lippitz, F. Kulzer, and M. Orrit, "Statistical Evaluation of Single Nano-Object Fluorescence," ChemPhysChem **6**, 770 (2005).

[220] M. Prummer, C. G. Hübner, B. Sick, B. Hecht, A. Renn, and U. P. Wild, "Single-Molecule Identification by Spectrally and Time-Resolved Fluorescence Detection," Anal. Chem. **72**, 443 (2000).

[221] R. Luchoswki, E. Matveeva, I. Gryczynski, E. Terpetschnig, L. Patsenker, G. Laczko, J. Borejdo, and Z. Gryczynski, "Single molecule studies of multiple-fluorophore labeled antibodies. Effect of homo-FRET on the number of photons available before photobleaching," Curr. Pharm. Biotechnol. **9**, 411 (2008).

[222] B. W. van der Meer, "Förster Theory," in *FRET - Förster Resonance Energy Transfer*, I. Medintz and N. Hildebrandt, eds. (Wiley-VCH, Weinheim, 2014).

[223] X.-L. Chu, T. Brenner, X.-W. Chen, Y. Ghosh, J. Hollingsworth, V. Sandoghdar, and S. Götzinger, "Experimental realization of an optical antenna for collecting 99% of photons from a quantum emitter," Optica **4**, 203 (2014).

[224] D. L. Dexter, "A Theory of Sensitized Luminescence in Solids," J. Chem. Phys. **21**, 836 (1951).

[225] D. Beljonne, C. Curutche, G. D. Scholes, and R. J. Silbey, "Beyond Förster Resonance Energy Transfer in Biological and Nanoscale Systems," J. Phys. Chem. B **113**, 6583 (2009).

[226] S. Tretiak and S. Mukamel, "Density Matrix Analysis and Simulation of Electronic Excitations in Conjugated and Aggregated Molecules," Chem. Rev. **102**, 3171 (2002).

[227] R. Métivier, F. Nolde, K. Müllen, and T. Basché, "Electronic Excitation Energy Transfer between Two Single Molecules Embedded in a Polymer Host," Phys. Rev. Lett. **98**, 047,802 (2007).

[228] S. Delaney and J. K. Barton, "Long-Range DNA Charge Transport," J. Org. Chem. **68**, 6475 (2003).

[229] C. S. Smith, N. Joseph, B. Rieger, and K. A. Lidke, "Fast, single-molecule localization that achieves theoretically minimum uncertainty," Nat. Methods **7**, 373 (2010).

[230] S. Stallinga and B. Rieger, "Accuracy of the Gaussian Point Spread Function model in 2D localization microscopy," Opt. Express **18**, 24,461 (2010).

[231] J. Engelhardt, J. Keller, P. Hoyer, M. Reuss, T. Staudt, and S. W. Hell, "Molecular Orientation Affects Localization Accuracy in Superresolution Far-Field Fluorescence Microscopy," Nano Lett. **11**, 209 (2011).

[232] M. D. Lew, M. P. Backlund, and W. E. Moerner, "Rotational Mobility of Single Molecules Affects Localization Accuracy in Super-Resolution Fluorescence Microscopy," Nano Lett. **13**, 3967 (2013).

[233] S. M. Kay, *Fundamentals of Statistical Signal Processing: Estimation Theory* (Prentice Hall, Upper Saddle River, 1993).

[234] D. R. Cox and D. V. Hinkley, *Theoretical Statistics* (CRC Press, Boca Raton, 1979).

[235] D. Zwillinger, *CRC Standard Mathematical Tables and Formulae* (CRC Press, Boca Raton, 1995).

[236] I. N. Bronstein and K. A. Semendyayev, *Handbook of Mathematics* (Springer, New York, 2004).

[237] A. C. Aitken, "On Least-squares and Linear Combinations of Observations," P. Roy. Soc. Edinb. A **55**, 42 (1934).

[238] P. Bordat and R. Brown, "Molecular mechanisms of photo-induced spectral diffusion of single terrylene molecules in p terphenyl," J. Chem. Phys. **116**, 229 (2002).

[239] S. W. Hell, "Microscopy and its focal switch," Nat. Methods **6**, 24 (2009).

[240] M. Heilemann, "Fluorescence microscopy beyond the diffraction limit." J. Biotechnol. **149**, 243 (2010).

[241] N. Bobroff, "Position measurement with a resolution and noise-limited instrument," Rev. Sci. Instrum. **57**, 1152 (1986).

[242] A. Van Oijen, J. Köhler, J. Schmidt, M. Müller, and G. Brakenhoff, "Far-field fluorescence microscopy beyond the diffraction limit," J. Opt. Soc. Am. A **16**, 909 (1999).

[243] R. Zondervan, F. Kulzer, M. A. Kolchenk, and M. Orrit, "Photobleaching of rhodamine 6G in poly (vinyl alcohol) at the ensemble and single-molecule levels," J. Phys. Chem. A **108**, 1657 (2004).

[244] A. Renn, J. Seelig, and V. Sandoghdar, "Oxygen-dependent photochemistry of fluorescent dyes studied at the single molecule level." Mol. Phys. **104**, 409 (2006).

[245] K. Pierce, R. Premont, and R. Lefkowitz, "Seven-transmembrane receptors," Nat. Rev. Mol. Cell Biol. **3**, 639 (2002).

[246] J. P. Overington, B. Al-Lazikani, and A. L. Hopkins, "How many drug targets are there?" Nat. Rev. Drug Discov. **5**, 993 (2006).

[247] M. Bouvier, "Oligomerization of G-protein-coupled transmitter receptors," Nat. Rev. Neurosci. **2**, 274 (2001).

[248] G. Milligan, "G protein-coupled receptor dimerization: function and ligand pharmacology," Mol. Pharmacol. **66**, 1 (2004).

[249] S. Ferre, V. Casado, L. Devi, M. Filizola, R. Jockers, M. Lohse, G. Milligan, J. Pin, and X. Guitart, "G protein-coupled receptor oligomerization revisited: functional and pharmacological perspectives," Pharmacol. Rev. **66**, 413 (2014).

[250] J.-M. Beaulieu and R. Gainetdinov, "The physiology, signaling, and pharmacology of dopamine receptors," Pharmacol. Rev. **63**, 182 (2011).

[251] K. Davis, R. Kahn, G. Ko, and M. Davidson, "Dopamine in schizophrenia: a review and reconceptualization," Am. J. Psychiatry **148**, 1474 (1991).

[252] A. Zhang, J. Neumeyer, and R. Baldessarini, "Recent progress in development of dopamine receptor subtype-selective agents: potential therapeutics for neurological and psychiatric disorders," Chem. Rev. **107**, 274 (2007).

[253] C. Pou, C. Mannoury la Cour, L. Stoddart, M. Millan, and G. Milligan, "Functional homomers and heteromers of dopamine D2L and D3 receptors co-exist at the cell surface," J. Biol. Chem. **287**, 8864 (2012).

[254] M. Wang, L. Pei, P. Fletcher, S. Kapur, P. Seeman, and F. Liu, "Schizophrenia, amphetamine-induced sensitized state and acute amphetamine exposure all show a common alteration: increased dopamine D2 receptor dimerization," Mol. Brain **3**, 25 (2010).

[255] J. A. Hern, A. H. Baig, G. I. Mashanov, B. Birdsall, J. E. T. Corrie, S. Lazareno, J. E. Molloy, and N. J. M. Birdsall, "Formation and dissociation of M1 muscarinic receptor dimers seen by total internal reflection fluorescence imaging of single molecules," Proc. Natl. Acad. Sci. USA **107**, 2693–2698 (2010).

[256] R. Kasai, K. Suzuki, E. Prossnitz, I. Koyama-Honda, C. Nakada, T. Fujiwara, and A. Kusumi, "Full characterization of GPCR monomer-dimer dynamic equilibrium by single molecule imaging," J. Cell Biol. **192**, 463 (2011).

[257] D. Calebiro, F. Rieken, J. Wagner, T. Sungkaworn, U. Zabel, A. Borzi, E. Cocucci, A. Zürn, and M. J. Lohse, "Single-molecule analysis of fluorescently labeled G-protein–coupled receptors reveals complexes with distinct dynamics and organization," Proc. Natl. Acad. Sci. USA **110**, 743 (2013).

[258] A. Juillerat, T. Gronemeyer, A. Keppler, S. Gendreizig, H. Pick, H. Vogel, and K. Johnsson, "Directed evolution of O6-alkylguanine-DNA alkyltransferase for efficient labeling of fusion proteins with small molecules in vivo," Chem. Biol. **10**, 313 (2003).

[259] C. Hiller, R. C. Kling, F. W. Heinemann, K. Meyer, H. Hübner, and P. Gmeiner, "Functionally Selective Dopamine D2/D3 Receptor Agonists Comprising an Enyne Moiety," J. Med. Chem. **56**, 5130–5141 (2013).

[260] G. Crivat and J. Taraska, "Imaging proteins inside cells with fluorescent tags," Trends Biotechnol. **30**, 8–16 (2012).

[261] N. Guex and M. Peitsch, "SWISS-MODEL and the Swiss-PdbViewer: an environment for comparative protein modeling," Electrophoresis **18**, 2714–2723 (1997).

[262] J. Huang, S. Chen, J. Zhang, and X. Huang, "Crystal structure of oligomeric β1-adrenergic G protein-coupled receptors in ligand-free basal state," Nat. Struct. Mol. Biol. **20**, 419–425 (2013).

[263] T. Dertinger, R. Colyer, G. Iyer, S. Weiss, and J. Enderlein, "Fast, background-free, 3D super-resolution optical fluctuation imaging (SOFI)," Proc. Natl. Acad. Sci. USA **106**, 22,287–22,292 (2009).

[264] T. Dertinger, M. Heilemann, R. Vogel, M. Sauer, and S. Weiss, "Superresolution Optical Fluctuation Imaging with Organic Dyes," Angew. Chem. Int. Ed. **49**, 9441–9443 (2010).

[265] S. Cox, E. Rosten, J. Monypenny, T. Jovanovic-Talisman, D. Burnette, J. Lippincott-Schwartz, G. Jones, and R. Heintzmann, "Bayesian localization microscopy reveals nanoscale podosome dynamics," Nat. Methods **9**, 195–200 (2012).

[266] R. M. Neal, "Probabilistic Inference Using Markov Chain Monte Carlo Methods," Techn. Rep. CRG-TR-93-1, Dept. of Computer Science, University of Toronto (1993).

[267] E. Rosten, G. Jones, and S. Cox, "ImageJ plug-in for Bayesian analysis of blinking and bleaching," Nat. Methods **10**, 97 (2013).

[268] P. K. Mattila and P. Lappalainen, "Filopodia: molecular architecture and cellular functions," Nat. Rev. Mol. Cell. Biol. **9**, 446 (2008).

[269] M. Heilemann, E. Margeat, R. Kasper, M. Sauer, and P. Tinnefeld, "Carbocyanine Dyes as Efficient Reversible Single-Molecule Optical Switch," J. Am. Chem. Soc. **127**, 3801 (2005).

[270] A. R. Faro, P. Carpentier, G. Jonasson, G. Pompidor, D. Arcizet, I. Demachy, and D. Bourgeois, "Low-temperature Chromophore Isomerization Reveals the Photoswitching Mechanism of the Fluorescent Protein Padron," J. Am. Chem. Soc. **133**, 16,362 (2011).

[271] D. T. Burnette, P. Sengupta, Y. Dai, J. Lippincott-Schwartz, and B. Kachar, "Bleaching/blinking assisted localization microscopy for superresolution imaging using standard fluorescent molecules," Proc. Natl. Acad. Sci. USA **108**, 21,081 (2011).

[272] P. D. Simonson, E. Rothenberg, and P. R. Selvin, "Single-Molecule-Based Super-Resolution Images in the Presence of Multiple Fluorophores," Nano Lett. **11**, 5090 (2011).

[273] M. A. Little and N. S. Jones, "Sparse Bayesian step-filtering for high-throughput analysis of molecular machine dynamics," in *Proc. IEEE Int. Conf. Acoust. Speech Signal Process.*, p. 4162 (2010).

[274] L. S. Churchman, H. Flyvbjerg, and J. A. Spudich, "A Non-Gaussian Distribution Quantifies Distances Measured with Fluorescence Localization Techniques," Biophys. J. **90**, 668 (2006).

[275] S. O. Rice, "Mathematical Analysis of Random Noise," Bell Syst. Tech. J. **24**, 46–156 (1945).

[276] F. Cichos, C. von Borczyskowski, and M. Orrit, "Power-law intermittency of single emitters," Curr. Opin. Colloid Interface Sci. **12**, 272 (2007).

[277] M. Lippitz, C. G. Hübner, T. Christ, H. Eichner, P. Bordat, A. Herrmann, K. Müllen, and T. Basché, "Coherent Electronic Coupling versus Localization in Individual Molecular Dimers," Phys. Rev. Lett. **92**, 103,001 (2004).

[278] A. Pertsinidis, K. Mukherjee, M. Sharma, Z. P. Pang, S. R. Park, Y. Zhang, A. T. Brunger, T. C. Südhof, and S. Chu, "Ultrahigh-resolution imaging reveals formation of neuronal SNARE/Munc18 complexes in situ," Proc. Natl. Acad. Sci. USA **110**, E2812–E2820 (2013).

[279] M. Scarselli, P. Annibale, C. Gerace, and A. Radenovic, "Enlightening G-protein-coupled receptors on the plasma membrane using super-resolution photoactivated localization microscopy," Biochem. Soc. Trans. **41**, 191 (2013).

[280] A. Borgia, P. M. Williams, and J. Clarke, "Single-molecule studies of protein folding," Annu. Rev. Biochem. **77**, 101 (2008).

[281] B. Schuler and H. Hofmann, "Single-molecule spectroscopy of protein folding dynamics-expanding scope and timescales," Curr. Opin. Struct. Biol. **23**, 36 (2013).

[282] H. Yuan, T. Xia, B. Schuler, and M. Orrit, "Temperature-cycle single-molecule FRET microscopy on polyprolines," Phys. Chem. Chem. Phys. **13**, 1762 (2011).

[283] D. L. Andrews, C. Curutchet, and G. D. Scholes, "Resonance energy transfer: Beyond the limits," Laser Photon. Rev. **123**, 114 (2011).

[284] F. Würthner, T. E. Kaiser, and C. R. Saha-Möller, "J-aggregates: from serendipitous discovery to supramolecular engineering of functional dye materials." Angew. Chem. Int. Ed. **50**, 3376 (2011).

[285] N. C. Dvornek, F. J. Sigworth, and H. D. Tagare, "SubspaceEM: A fast maximum-a-posteriori algorithm for cryo-EM single particle reconstruction," J. Struct. Biol. **190**, 200–214 (2015).

[286] G. Harauz and M. van Heel, "Exact filters for general geometry three dimensional reconstruction," Optik **73**, 146–156 (1986).

[287] M. van Heel and M. Schatz, "Fourier shell correlation threshold criteria," J. Struct. Biol. **151**, 250–262 (2005).

[288] K. Kolmakov, V. N. Belov, C. A. Wurm, B. Harke, M. Leutenegger, C. Eggeling, and S. W. Hell, "A Versatile Route to Red-Emitting Carbopyronine Dyes for Optical Microscopy and Nanoscopy," Eur. J. Org. Chem. **2010**, 3593–3610 (2010).

[289] E. Pettersen, T. Goddard, C. Huang, G. Couch, D. Greenblatt, E. Meng, and T. Ferrin, "UCSF Chimera–a visualization system for exploratory research and analysis," J. Comput. Chem. **25**, 1605 (2004).

[290] M. Sevvana, V. Vijayan, M. Zweckstetter, S. Reinelt, D. R. Madden, R. Herbst-Irmer, G. M. Sheldrick, M. Bott, C. Griesinger, and S. Becker, "A ligand-induced switch in the periplasmic domain of sensor histidine kinase CitA," J. Mol. Biol. **377**, 512 (2008).

[291] S. Kirkpatrick, C. D. Gelatt Jr, and M. P. Vecchi, "Optimization by Simulated Annealing," Science **220**, 671 (1983).

[292] Y. D. Tsvetkov and Y. A. Grishin, "Techniques for EPR spectroscopy of pulsed electron double resonance (PELDOR): A review," Instrum. Exp. Tech. **52**, 615 (2009).

[293] M. P. Backlund, M. D. Lew, A. S. Backer, S. J. Sahl, G. Grover, A. Agrawal, R. Piestun, and W. E. Moerner, "Simultaneous, accurate measurement of the 3D position and orientation of single molecules," Proc. Natl. Acad. Sci. USA **109**, 19,087 (2012).

[294] M. Lang, T. D. Milster, T. Minamitani, G. Borek, and D. Brown, "Fabrication and Characterization of Sub-100 µm Diameter Gallium Phosphide Solid Immersion Lens Arrays," Jpn. J. Appl. Phys. **44**, 3385 (2005).

[295] K. G. Lee, X. Chen, H. Eghlidi, P. Kukura, R. Lettow, A. Renn, V. Sandoghdar, and S. Götzinger, "A planar dielectric antenna for directional single-photon emission and near-unity collection efficiency," Nat. Photonics **5**, 166 (2011).

[296] J. W. Taraska, M. C. Puljung, N. B. Olivier, G. E. Flynn, and W. N. Zagotta, "Mapping the structure and conformational movements of proteins with transition metal ion FRET," Nat. Methods **6**, 532 (2009).

[297] G. Y. Shevelev, O. A. Krumkacheva, A. A. Lomzov, A. A. Kuzhelev, D. V. Trukhin, O. Y. Rogozhnikova, V. M. Tormyshev, D. V. Pyshnyi, M. V. Fedin, and E. G. Bagryanskaya, "Triarylmethyl Labels: Toward Improving the Accuracy of EPR Nanoscale Distance Measurements in DNAs," J. Phys. Chem. B **119**, 13,641 (2015).

[298] A. Muschielok, J. Andrecka, A. Jawhari, F. Brückner, P. Cramer, and J. Michaelis, "A nanopositioning system for macromolecular structural analysis," Nat. Methods **5**, 965 (2008).

[299] M. Lippitz, W. Erker, H. Decker, K. E. van Holde, and T. Basché, "Two-photon excitation microscopy of tryptophan-containing proteins," Proc. Natl. Acad. Sci. USA **99**, 2772 (2002).

[300] A. Löschberger, S. van de Linde, M.-C. Dabauvalle, B. Rieger, M. Heilemann, G. Krohne, and M. Sauer, "Super-resolution imaging visualizes the eightfold symmetry of gp210 proteins around the nuclear pore complex and resolves the central channel with nanometer resolution," J. Cell Sci. **125**, 570 (2012).

[301] M. Lehmann, B. Gottschalk, D. Puchkov, P. Schmieder, S. Schwagerus, C. P. R. Hackenberger, V. Haucke, and J. Schmoranzer, "Multicolor Caged dSTORM Resolves the Ultrastructure of Synaptic Vesicles in the Brain," Angew. Chem. Int. Ed. **54**, 13,230 (2015).

[302] R. Zondervan, F. Kulzer, H. van der Meer, J. A. Disselhorst, and M. Orrit, "Laser-Driven Microsecond Temperature Cycles Analyzed by Fluorescence Polarization Microscopy," Biophys. J. **90**, 2958 (2006).

[303] C. K. E. Bleck, A. Merz, M. G. Gutierrez, P. Walther, J. Dubochet, B. Zuber, and G. Griffiths, "Comparison of different methods for thin section EM analysis of Mycobacterium smegmatis," J. Microsc. **237**, 23 (2010).

[304] U. Schnell, F. Dijk, K. A. Sjollema, and B. N. Giepmans, "Immunolabeling artifacts and the need for live-cell imaging," Nat. Methods **9**, 152 (2012).

[305] B. Weinhausen, O. Saldanha, R. N. Wilke, C. Dammann, M. Priebe, M. Burghammer, M. Sprung, and S. Koester, "Scanning X-ray nanodiffraction on living eukaryotic cells in microfluidic environments," Phys. Rev. Lett. **112**, 088,102 (2014).

[306] V. Westphal, S. O. Rizzoli, M. A. Lauterbach, D. Kamin, R. Jahn, and S. W. Hell, "Video-Rate Far-Field Optical Nanoscopy Dissects Synaptic Vesicle Movement," Science **320**, 246 (2008).

[307] S. A. Jones, S.-H. Shim, J. He, and X. Zhuang, "Fast, three-dimensional super-resolution imaging of live cells," Nat. Methods **8**, 499 (2011).

[308] F. Huang, T. M. P. Hartwich, F. E. Rivera-Molina, Y. Lin, W. C. Duim, J. J. Long, P. D. Uchil, J. R. Myers, M. A. Baird, W. Mothes, M. W. Davidson, D. Toomre, and J. Bewersdorf, "Video-rate nanoscopy using sCMOS camera–specific single-molecule localization algorithms," Nat. Methods **10**, 653 (2013).

[309] I. Hurbain and M. Sachse, "The future is cold: cryo-preparation methods for transmission electron microscopy of cells," Biol. Cell **103**, 405 (2011).

[310] G. Schneider, P. Guttmann, S. Rehbein, S. Werner, and R. Follath, "Cryo X-ray microscope with flat sample geometry for correlative fluorescence and nanoscale tomographic imaging," J. Struct. Biol. **177**, 212 (2012).

[311] G. Lukinavicius, L. Reymond, E. D'Este, A. Masharina, F. Göttfert, H. Ta, A. Güther, M. Fournier, S. Rizzo, H. Waldmann, C. Blaukopf, C. Sommer, D. W. Gerlich, H.-D. Arndt, S. W. Hell, and K. Johnsson, "Fluorogenic probes for live-cell imaging of the cytoskeleton," Nat. Methods **11**, 731 (2014).

[312] M. Lakadamyali and M. P. Cosma, "Advanced microscopy methods for visualizing chromatin structure," FEBS Lett. **589**, 3023 (2015).

[313] P. R. Langer-Safer, M. Levine, and D. C. Ward, "Immunological method for mapping genes on Drosophila polytene chromosomes," Proc. Natl. Acad. Sci. USA **79**, 4381 (1982).

[314] R. Amann and B. M. Fuchs, "Single-cell identification in microbial communities by improved fluorescence in situ hybridization techniques," Nature Rev. Microbiol. **6**, 339 (2008).

[315] J. Dekker, K. Rippe, M. Dekker, and N. Kleckner, "Capturing Chromosome Conformation," Science **295**, 1306 (2002).

[316] M. Jinek, K. Chylinski, I. Fonfara, M. Hauer, J. A. Doudna, and E. Charpentier, "A Programmable Dual-RNA–Guided DNA Endonuclease in Adaptive Bacterial Immunity," Science **337**, 816 (2012).

[317] A. N. Boettiger, B. Bintu, J. R. Moffitt, S. Wang, B. J. Beliveau, G. Fudenberg, M. Imakaev, L. A. Mirny, C.-T. Wu, and X. Zhuang, "Super-resolution imaging reveals distinct chromatin folding for different epigenetic states," Nature **529**, 418 (2016).

Fun facts about this work

- Liquid helium consumption (in liters) [1]: 7,500
- Coffee consumption (in cups) [2]: 4,200
- Number of Hungry Thursdays™(including Hungry Wednesdays™) [3]: 27
- Miles traveled for conferences and seminars [4]: 101,400
- Number of postcards sent [5]: 34
- Amount of experimental data generated (in GB) [6]: 1,600
- Lines of source code [7]: 70,100
- Total points won playing Skat [8]: 18,675
- Number of China gadgets bought [9]: > 35
- Time to complete this work (in hours) [10]: 14,500

[1] Mean value calculated from the logbook of the liquid helium filling and the gas counter reading, assuming 1 l of liquid helium correspond to 0.76 m^3 of gas.

[2] Extrapolated from the average consumption per month from the time period after the introduction of the *stroke list*.

[3] Computed from detailed recordings.

[4] Approximated from list of destinations.

[5] Counted on the wall in the *green room*.

[6] Output of `du -csh --block-size=1G .` for the experimental data directory.

[7] Output of `find . -name '*.m' | xargs wc -l` for the source code directory.

[8] Computed from detailed recordings.

[9] Approximated by number of emails from eBay in the inbox.

[10] Conservative estimate based on a five-day work week.

Acknowledgements

Finally, I would like to thank and acknowledge the people whose help, support and encouragement allowed me to complete this work. I have had the sincere privilege and utmost pleasure to work and spend time with the most exceptional and brilliant colleagues from all over the world.

First and foremost, I am indebted to my doctoral advisor and mentor Vahid for his continuous support and advice. He is an accomplished master in motivation and optimism, and his enthusiasm and creativity as well as his warm and friendly personality make his lab an inspiring place to do research. Another one of his superpowers is to make time for personal advice and chit-chat even when hopelessly swamped with appointments and manuscripts on the one day between coming back from the US and leaving for Japan.

At this point I should list all the members of the nano-optics group, be it at the Max Planck Institute in Erlangen or back at ETH in Zürich. Since I do not want to forget anyone, I choose to not list their names here. You know that I mean you. Especially as an experimentalist, one often relies on each other—both in the lab and in scientific discussions over coffee or tea in the *green room*. I have had a wonderful time with my labmates, officemates and all the other *nanos*, both at work and at various other occasions such as house-warming parties, Hungry Thursdays™, Bergkirchweih, barbecues, Skat evenings in the pub, badminton, soccer, golf, laser tag, kart racing, costume parties, group seminars, ski workshops, Starcraft sessions, and many more.

一纳米是非常特殊的。我感谢命运满足露晞。你的快乐激励着我。我幸福因为你的爱。我全心全意的爱你。

Zuletzt möchte ich meinen Eltern Johanna und Josef danken für ihre Liebe, Vertrauen und Unterstützung während meines ganzen Lebens. Sie haben mir immer die Freiheit gegeben, das zu tun, was ich möchte. Ihnen widme ich diese Dissertation.